THE 4-DIMENSIONAL AFTERLIFE AND OUR PLANETARY CHANGES

By

Fergus Davison

MAPLE
PUBLISHERS

THE 4-DIMENSIONAL AFTERLIFE AND OUR PLANETARY CHANGES

Author: Fergus Davison

Copyright © Fergus Davison (2025)

The right of Fergus Davison to be identified as author of this work has been asserted by the author in accordance with section 77 and 78 of the Copyright, Designs and Patents Act 1988.

First Published in 2025

ISBN 978-1-83538-835-8 (Ppaerback)
    978-1-83538-903-4 (E-Book)

Cover Design and Book Layout by:
    Maple Publishers
    www.maplepublishers.com

Published by:
    Maple Publishers
    Fairbourne Drive, Atterbury,
    Milton Keynes,
    MK10 9RG, UK
    www.maplepublishers.com

The views expressed in this work are solely those of the author and do not reflect the opinions of Publishers, and the Publisher hereby disclaims any responsibility for them. This book should not be used as a substitute for the advice of a competent authority, admitted or authorized to advise on the subjects covered.

A CIP catalogue record for this title is available from the British Library.

All rights reserved. No part of this book may be reproduced or translated by any form or by any means, electronic or mechanical, including photocopying, recording or by any information storage and retrieval system without written permission from the author.

It is possible that our 3-dimensional universe encloses a spherical volume containing another spatial dimension. While inhabiting a physical body on Earth we perceive this extra dimension as time but after death we return to this universe and live in a 4-dimensional afterlife without time or change as we understand it here.

CONTENTS

Forward 1

Introduction 4

The Shape of The Universe 9

    Some Points of Numerical Interest 13

Figure 1. Space-Time Diagram 15

Inside The Hypersphere 16

The Soul (Astral or Real Body) 18

Evidence For An Afterlife 20

Out of Body Experience 22

Figure 2. Afterlife Spheres Around The Earth 26

How The Metaphysical Dimensions Work 27

Our Metaphysical Origin 35

The Lower Metaphysical Region 37

Life Forms in the Lower Metaphysical Regions 40

Cryptids 42

The Occult 49

    Pest Control 51

Astral Intrusions In The Physical Dimension 54

    Time Slips 54

Figure 3. Time Slip Diagram 57

Predetermination 58

Memory 64

The Judgement 69

    Synchronicity: Do Prayers Work? 71

THE COMING PLANETARY CHANGES	74
Apocalyptic Possibilities	74
The Dead Man's Fuse	79
Avoiding Armageddon	81
Extraterrestrials (ETs)	84
The Greys	86
Antigravity	95
Humans, Humanoids And Brains	99
Figure 4. Skull Evolution	103
Disclosure	104
The Sirius Disclosure Project	116
Ancient Aliens And Our Evolution	121
Summary	125
Bibliography	131

Forward

I have spent most of my professional life in medical scientific research, predominantly in genetics, molecular biology and virology. This scientific background has enabled me to appreciate the importance of the scientific method where hard data is crucially important. I have learned how scientists think and how scientific methodology is followed. Experimental evidence must be obtained which is repeatable so that a scientific hypothesis can be tested and confirmed. And ideally, to quote the famous scientific philosopher Karl Popper, such a hypothesis must be falsifiable should contradictory evidence become available. This approach to science has led to amazing discoveries and provided humanity with an abundance of technology to make our existence easier and happier in so many areas. Especially in the fields of medicine, communication, transport, labour-saving devices and entertainment.

However, conventional science only deals with the physical universe and the physical laws that exist here. As yet, conventional science sees no real evidence of a non-physical, or metaphysical universe, especially as nothing can be measured or routinely observed under controlled laboratory conditions. If any evidence does appear it becomes almost impossible for this new knowledge to be adequately peer-reviewed in our scientific climate and is frequently resisted by the guardians of the establishment. There *are* a few measurable phenomena which cause difficulties in comprehension, on the quantum scale, like the wave-particle duality of photons and entangled atomic particles communicating instantaneously, regardless of their separation, as in Einstein's intellectual problem with "spooky action at a distance". Theoretical physicists are happy to address these contradictions as they are clearly repeatable and measurable. But when it comes to the science of parapsychology and investigating the paranormal then conventional science becomes very cautious and dismissive. The demand for hard evidence is so rigid that any alternative sources of information cannot and must not be considered. Any scientist proclaiming belief in such phenomenon would have their professional reputation ruined. We have had some modest paranormal investigations eg. a few university psychology departments will obtain test subjects to try and modify the random fall of dice or guess the sequence of dealt playing cards using the power of mind. If there is a statistical deviation from the chance result, then something unexplained must be occurring. But this is only the small stuff. Have they found anyone who can perform sustained levitation in a laboratory or read someone else's mind accurately like a book or cure diseases instantly? Then no.

The main problem with non-conventional science is that it is unpredictable and not repeatable. I have had to suspend my allegiance to the orthodox scientific method and accept that there is a hidden part of our existence and our physical universe that operates with different rules. This means I am open to the charge of just promoting pseudo-science, like trying to prove the existence of God in a laboratory. My simple answer to this is, wait and see and I am very optimistic. I do not know when this will happen, but humanity is reaching a critical point so I suspect not much longer. I believe humanity is being deliberately blocked from an awareness of the non-physical universe or what is sometimes called revealed knowledge as opposed to empirical knowledge which in principle everyone can access through the scientific method. Possibly there is a particular cosmic reason for this and there is a due process that must take place on all inhabited planets. I suspect that all intelligent beings are obliged to do a certain amount of "time" before we are mature enough for revealed knowledge to become available to us. This is frustrating to many people, including myself, who realise there is something there but we can only perceive glimpses. Alternatively, I might be accused of promoting pseudo-religion. However, there are a number of places in our religious texts, especially the Bible, which highlight some of the points I am trying to make. I am happy to cherry-pick these but generally I think religious scripture, certainly within the Abrahamic faiths, has caused more human strife rather than comfort over the ages.

The basic hypothesis of this brief book can be summed up in a single sentence. The universe is an expanding 4-dimensional sphere brought into being by a Superintelligence. However, the implications and consequences of this creation are vast and currently beyond our understanding. Humanity's lack of awareness of the hidden dimension in our physical life over thousands of years is a problem for mankind because it leaves a vast gap in the understanding of our existence and purpose. Hopefully it will soon be revealed. What I have written here is meant to be a brief handbook for what we should expect soon and why it is happening.

There are many references to other authors and fairly generalized subject matters in the text which I have not individually referenced. This is because the internet is in such common usage now that a simple online search will provide a wealth of follow-up information. For example, Wikipedia provides in-depth details on subjects like cryptozoology and the CVs of well-known scientists. However, in some special cases I have referenced a particular book or text when this seems more helpful.

"The day science begins to study non-physical phenomena; it will make more progress in one decade than in all the previous centuries of its existence." Nikola Tesla.

Introduction

A lot of the big questions in life few people seriously attempt to answer, like why are we here? How does the cosmos work, especially, is there a purpose for us all? What I would like to set out here is a theory of how our lives in the physical dimension of the universe may relate to an existence that continues after our physical bodies have died. Once you have made the decision that there is some sort of grand system at work for this purpose, then how does it work? Of course, much of this will be speculation as getting hard data is extremely difficult and any research in this area is normally greeted with a good measure of scorn from those in orthodox science. Also, proof of life after death will provide plausible explanations for practically all paranormal phenomena, many of which at present seem unconnected. Lack of evidence is a big problem and confirmed skeptics like Richard Dawkins, the Oxford University academic and writer, normally win the argument here so if you are one of these there is little point reading on. Read his book *The God Delusion* (1). Fortunately, there are some scientists but sadly only a few, who are more enlightened like Rupert Sheldrake and have taken the opposite view. He has written much on the subject of a conscious universe, especially in his book *The Science Delusion* (2). However, there is a mountain of anecdotal evidence for human existence continuing after physical death. Firstly, there are the "spooky" phenomena. Apparitions, hauntings and poltergeists have been reported on a grand scale around the world ever since recorded history began. Secondly, there are a small number of genuine mediums who seem very reliable and provide quite astonishingly accurate information to living relatives of the deceased. Thirdly, there are many well- documented descriptions of out-of-body experiences in which individuals have described a non-physical existence distinct from that on earth. Fourthly, some people claim types of religious experience in which they receive an indescribable feeling of spiritual enlightenment and need no further convincing that they have a spiritual dimension to their existence. Finally, I should include those who claim to have close encounters with extraterrestrials (ETs). They are often told that their physical bodies also have a non-physical component and their lives are courtesy of a Supreme Being. But conventional thought, ie. orthodox science will have none of this, in spite of the overwhelming body of non-testable evidence. A good example of scientists willfully ignoring anomalies took place during the time of Galileo in the 16th century. The scientific establishment at that time included the Church and when Galileo said he could demonstrate that the Earth was not the centre of the Solar System, the church leaders would bluntly refuse to look through his telescope, stating this was not necessary because they already knew the truth. A similar situation occurs with our current official scientists who for example, absolutely refuse to investigate peculiar phenomenon like crop circles; the more sophisticated of which are clearly made not using technology of terrestrial origin, or hoaxing by a couple of bored guys down the pub in Wiltshire.

There comes a point when refusing to acknowledge the reality of paranormal data becomes quite foolish. I accept that from time to time an elaborate hoax may be staged but not on a scale to explain away so much of the evidence. What is responsible for such paranormal phenomena? Some of the suggested conventional explanations for these events are more implausible than the possibility of a continuing non-physical existence. However, death is a rather taboo subject and most of us would prefer to change the subject; come on lighten up, whose round is it anyway? We are probably the only species on the planet that understands the concept of death and it is surprising that we don't look into it bit more. We are immersed in our daily business and problems; death for most of the time is a remote issue. On a universal time scale our physical lives are a tiny blip. The universe is almost 14 billion years old and our life span is about 3 score years and 10. If there is life after death, that is an awful lot of time *not* in a physical existence. This may include a number of reincarnations, should these be factored in. Maybe we should all keep this in mind a little more often as our real existence includes an eternity. Many of the world's religions have tried to address these issues although rather clumsily, violently and unsuccessfully. This is because, I believe, a veil that separates this physical world from the "other side" is currently very firmly in place. Why this veil is currently hiding the non-physical universe from us is a very profound and perplexing question. There have been many generations of humanity over thousands of years that have had this knowledge withheld and if we were aware, it would have made vast transformational changes to mankind's understanding of its purpose and progress. As a member of the Society for Psychical Research, Tony Cornell, pointed out in 1994. There are over 10000 *deceased* members of our Society but not one has bothered to stay in touch. Sometimes on this planet we are given enlightened individuals who incarnate here to guide us. When Jesus was at the end of his life he said: forgive them Father for they know not what they do. This seems so true but is it somehow our fault? It is certainly humanity's ignorance, but why is our ignorance so deeply embedded? Possibly when enough people on this planet accept there is an afterlife dimension to our existence, and we attempt to tune into it, then a planetary awareness will automatically follow. It will be like receiving an additional sense to the five we already have.

Sometimes we already call it a sixth sense. But as to how and when we get such "proof" then it seems we must wait.

However, these changes may come in the near future under circumstances that many Christians call the second coming. These changes are likely to reveal incredible knowledge and abilities which we will have at our disposal and in effect, would make our planet the Garden of Eden. The fact that there is so little obvious communication from those who have left the physical world is a serious problem and a mystery but there may be some good reasons for this. The two main possibilities are: a lack of desire to do so, or more likely an inability to do so. Should a small number of people have privileged access to afterlife information they could easily change the course of world history. If a malevolent individual with this sort of access should emerge it may provide an explanation for the predicted Biblical antichrist. Therefore, access is currently being restricted to occasional snippets of a more personal nature. For those who follow spiritualist teachings, like readings from The Joseph Foundation, claim that too much awareness of our past lives and future potential will short circuit our spiritual growth and development. All of us are living in the physical plane are here to learn, and the choices we make and the journey we take through life, is in many ways more important than the destination, which is in effect a return to our home.

There are many eminent people, especially in science, who go to extraordinary lengths to avoid invoking the possibility of a Creator. In general, science has a big problem with the concept of God and the intellectual acrobatics that some of them use to explain our existence in a perfectly balanced universe are often more implausible than the simple possibility that this Creator may exist. Interestingly, we have another group who only believe in the absolute truth of scripture and tell us that the Earth was created around 4000 BC in seven days by the mighty hand of a Divine Being coming down through the clouds. It is baffling that so many people with academic and scholarly backgrounds, even in science, prefer this version of creation. They choose to ignore all the hard evidence around us, much of it in the ground and in outer space. Although, I am sure there is a Creator, it was not and could not be achieved this way. In order for the Universe to be self-sustaining and not requiring constant adjustment it had to assemble naturally and gradually from the bottom up, over an enormous period of time, ie. 14 billion years. This type of creation means our universe can function independently, without constant tinkering from the Divine source being required. It is a concept often referred to as Intelligent Design.

There is very little agreement among those who discuss the after- life dimension regarding what to call the place we go to after leaving Earth. This dimension is our natural and final home, so referring to it as an afterlife is looking at it from the wrong perspective because it is our eventual real existence. It would make more sense to refer to the Earth plane or physical universe as the transient-life dimension. There are only two absolute dimensions of existence in Creation: the *physical* universe containing galaxies, stars and planets based on mass and energy, where we reside now and the non-physical. So, for the sake of simplicity, I am calling the unseen non-physical universe the *metaphysical*. The Greek prefix *meta* means beyond and should include all the many higher levels and sub-dimensions existing in the afterlife realms, these places being our eventual natural home. It can also define the region of paranormal experiences and unusual phenomena that exist outside our normal senses, including what we often call the supernatural.

The Shape of The Universe

Many of the fundamental questions on the origin and structure of the universe have stumped all those with the highest IQs in the business. Astrophysicists, cosmologists and mathematicians have got quite a lot of it sorted out but there are a number of fundamental issues they can't agree on including how many dimensions exist in the universe. There are the obvious 4 dimensions: 3 spatial and one of time, which are apparent to us all on a large scale, as we go about our life and regulate the movement of planets, stars and galaxies. To include the micro (quantum) scale, that involves the behaviour of sub-atomic particles, the experts tell us we may need a lot more dimensions, maybe even eleven, seven of which we do not perceive as they are so small.

One possibility is that the shape of the universe is a 4-dimensional sphere, sometimes called a hypersphere. This is clearly very difficult for us to comprehend as we are 3 dimensional beings but the concept is not new and has been suggested by several scientists of the past notably Einstein and Alexander Friedmann. Other peculiar shapes have also been postulated eg. saddles, cones, doughnuts and horns but the sphere is Nature's preferred structure and the most energy stable and provides the maximum volume for a minimal surface area. If an explosion occurred in a void the products would radiate outwards in a spherical shell around its source (ie. the "Big Bang"). However, the universe does behave in principle very like an inflating balloon but with an extra dimension present. Scientists have known for some time that the universe is expanding in a way that suggests it did so, starting from an initial position, then rather like the surface of a balloon all points on the surface are moving away from each other. Precisely what would happen if you marked several points on the surface of an inflating balloon. Cosmologists have observed that the rate of expansion is accelerating and speculate this may be due to something they call dark energy. There is some evidence that space is curved, ie. if you started a journey in a straight line and had enough time you would eventually arrive back where you started, as you would with a journey around a globe or a balloon's surface. If the universe is flat then it could be described as "infinite" therefore, going on forever. There are some problems with the concept of the universe being infinite. Cosmologists claim they can scroll back in time to a point of origin, or the Big Bang, around 14 billion years ago. If there is a point of origin and a defined time has elapsed since that point, then there must be a leading "edge", in this case the spherical 4-dimensional surface.

Furthermore, the universe is observed to be still expanding. How can "infinity" be expanding if it is already infinite in size? Conversely, if the universe is expanding then it must have been smaller in the past, so could infinity therefore be described at some earlier point as small? There is an inherent contradiction here. Theoretically if the universe were not expanding then it could be infinite. But if the universe is expanding then it must have an edge and therefore will be curved rather than flat. There is currently some disagreement amongst cosmologists on this issue, some claim space may be "flat" but some recent research has hinted at positive curvature which would suggest a sphere is a possibility, *Nature Astronomy* (3) and more recently from Professor Gastanaga at Portsmouth University (4). The point of origin of the big bang is likely to be at the centre of this hypersphere and therefore "perpendicular" to our 3D physical universe; we cannot travel there but in the 4th dimension we could. An expanding universe is unlikely to be flat because it must have an edge. A 4-dimensional (4D) sphere also very neatly gives our 4th dimension, or time, a linear quality in keeping with the other 3 linear, spatial dimensions in which we exist. Time of course is the radius of the 4D sphere. We do not perceive time spatially as we exist on the "surface" of this 4D sphere but we feel its effects and as the sphere expands the radius lengthens at a corresponding rate. In simple geometric terms the radius and circumference of a sphere are directly proportional regardless of the number of dimensions. In some ways time for us is an illusion, as we live in an everlasting present at the sphere's surface, or an alternative analogy would be that present time exists as the crest of a perpetual wave; it is a continuously moving instant in time. We use the term time to describe the process of change and it appears to have direction according to the physicist, Stephen Hawking (5). A possible consequence might be that if the universe stopped expanding then time would also stop. In fact, this universal expansion may determine that everything is perpetually moving in 4 dimensions and if it ceased the universe would become "frozen." Another consequence would be that there is no possibility of time travel as we cannot travel to a place (in the past or future) that does not exist. There is only one surface on the 4D sphere and that is the present (see space time diagram below). In theory it should be possible to measure time in feet and inches if we were able to determine the circumference of the universe, as a simple calculation would relate this to a value for the sphere's radius (ie. time). However, the *rate* of time passing can be altered by space/time warping. It has been shown experimentally that the passage of time can be slowed by gravity (or mass) confirming Einstein's theory of relativity. If our brains are prepared to comprehend the concept of a 3-dimensional surface then space, time and mass can be neatly described by visualizing their effect on the geometry of such a surface.

Our universe seems to be in what cosmologists describe as a "goldilocks" zone. If some of its basic parameters were even slightly altered it could not function in a way that is sustainable, certainly for life to evolve. These include the strength of forces that hold atoms together and the way different forces balance, for example between nuclear forces and electromagnetism. There have been other interesting observations from some scientific laboratories including a group led by Professor Webb (6) at Sydney University that the universe is non-isotropic or non-homogenous, which means there are slight variations in density and structure. They discerned that there were differences in the Planck constant when viewing the universe in different directions. It appears the universe may have a directionality that could be described as dipolar ie. having a north and south pole. This is tentative evidence of a spherical shape and is presumably rotating as well as expanding, but how could we test this? So how does such a shape relate to locating the afterlife? Clearly this is a difficult one to answer as the question of location for an afterlife has occupied the minds of religious leaders for eons. Heaven, Hell or Purgatory is a place obviously neither up nor down in the directional sense. It would seem that it is all around us and occasionally may interweave in a quasi-physical way with our own world and universe. I favour the simplest answer which is the existence of another spatial dimension that is hidden from our own. A further consequence of this hypothesis would be that all of Creation, ie. the physical and metaphysical universes evolved from the same time point. The "Big Bang" created Heaven and Earth together. Below I have described a possible shape for the universe that would account for this.

A century ago, an American mathematician, William A. Granville published *The Fourth Dimension and the Bible* (7). He believed that the purity of mathematics was an essence of truth from God. He speculated that the higher places and Heaven could only be accessed through the 4th dimension and claimed that we on Earth would perceive this 4th dimension as time. He believed that many of the miracles achieved by Jesus like creating large amounts of matter from nothing require the presence of this extra dimension, eg. the feeding of the 5000. Other miracles like Jesus passing through solid matter and calming storms could be attributed to his manipulation of an extra dimension. The intersection of 4 dimensional objects into our 3-dimensional world would seem incomprehensible to us as 3 dimensional beings and could be described as miraculous. The explanation of a geometric higher dimension interacting with a lower one was described by Edwin Abbot in his novel *Flatland* (8).

Some Points of Numerical Interest

In theory, the size of all God's creation can be calculated if we know the radius (R) of the universe. Cosmologists tell us that the age of the universe is approximately 14 billion years. If the elongation of this radius has been occurring at the speed of light, then that gives a value of 14 billion (14×10^9) light years for R. The formula for the volume of a hypersphere with 4 dimensions = $\pi^2/2 \times R^4$.

So, if a hypersphere has a hypervolume of $\pi^2/2 \times R^4$ and one light year is approximately 9.5×10^{12} kilometres.

Then substituting this value for R, the volume of the 4D universe = $\pi^2/2 \times (14 \times 10^9 \times 9.5 \times 10^{12})^4$. Which comes to about 6×10^{92} volumetric kilometres.

This equation results in a fantastically large number and is probably a bit meaningless in reality. But this result is likely to be an underestimate because cosmologists tell us that there are galaxies so far away that light from them will never reach us. This is because they are receding from us at a rate that exceeds the speed of light and the rate is accelerating. Or more accurately the space between us and those distant galaxies is expanding faster than light speed. That distance represents part of the circumference or perimeter of our universe which is equal to $2\pi R$. There is a linear relationship between a radius and the circumference of a sphere or circle so they must lengthen at the same rate; if the circumference is accelerating, then so is the radius. If the radius of the universe is extending at a speed greater than light and that rate is increasing, this would suggest that time is gradually speeding up also. The significance of this possibility is not clear. On a slightly frivolous note; I think we all feel that time passes more quickly as we get older.

Our 3D physical universe will have a "surface" volume of $2\pi^2R^3$. So, substituting the value for R as above we get a value of 4.6×10^{70} cubic kilometres. If we substitute for the value for R in $2\pi R$ then the circumference of our 4D universe is about 90 billion light years. This maybe about right as astronomers say that any object more than 46 billion light years away is receding at greater than light speed and will always be beyond our visible horizon. This is the limit of observable information. However, the fabric of space- time can expand faster than the speed of light over large distances according to cosmologists. The size of the physical universe is unbelievably vast. The ratio of the volume of a sphere to its surface area is enormous and this would also apply to a hypersphere. So, there would be no shortage of space in the place where we may all eventually find ourselves. The expansion within a volume is rather like observing raisins moving further apart in a rising cake. And cosmologists use this analogy when viewing galaxies in our expanding universe on a cosmic scale. But our 3D universe is only the "skin" on a 4D sphere. With the consequences of an extra spatial dimension, we must consider this process on a different scale. So, experiencing the expansion of space within a large expanding balloon is very different from the perspective of being at its surface. In other words, the increasing radius (or time) of the sphere is hardly perceived from inside the sphere, only at its surface. "Our" time becomes "their" 4th spatial dimension. See space-time diagram on the next page.

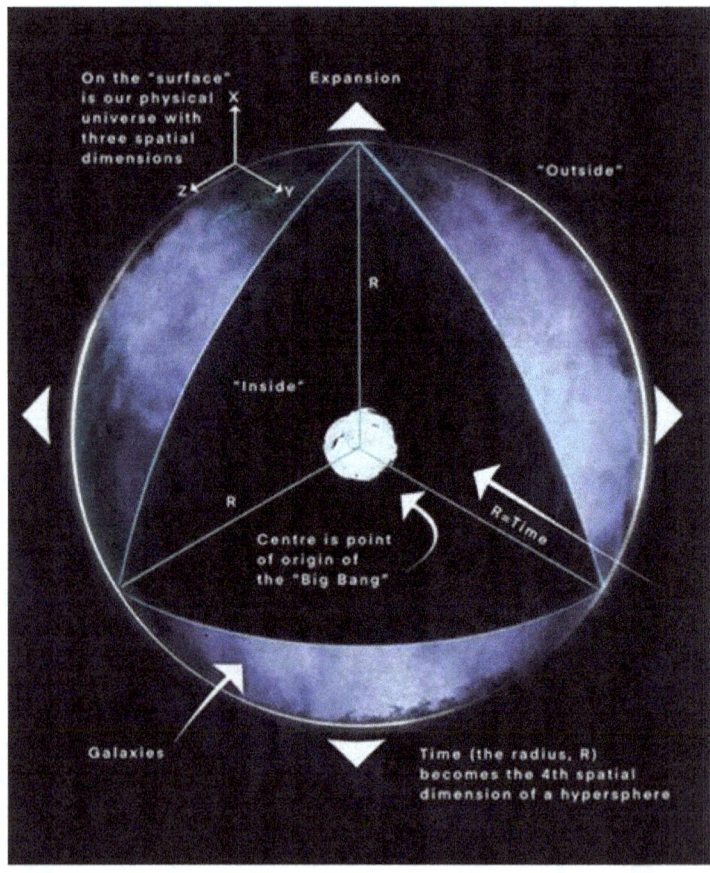

Figure 1. Space-Time Diagram

Inside The Hypersphere

I suggest that our 3-dimensional universe encloses a volume containing another spatial dimension as represented in the diagram above. So, does the hypersphere have an "outside" and "inside"? I maintain that outside this 4D sphere there is nothing; like there was nothing before the big bang. This is another concept we struggle with as we perceive our lives in terms of a linear time and space. The possibility that there may be a place with no time, space or matter is something we must just accept as our 3- dimensional brains fail here. However, *inside* the 4D spherical universe is another matter. What happens when we die and leave our 3-dimensional bodies? As a believer in a continuing existence after the death of our physical bodies, where is it that we go? Our 3-dimensional material bodies can only exist at the 3D surface of a 4D spherical universe so on physical death we can move to various locations within this 4D sphere and away from the surface. We may be 4D beings contained in a 3D shell for relatively short periods. It seems that all living organisms that have any individual awareness, including the most simple, will have an existence in a metaphysical region that reflects its prior existence in the physical world. For example, when a fish dies here it will automatically reappear in an appropriate region with an ocean. Where is God? Is another rather rhetorical question that people often ask whether as believers or skeptics, as he does seem hard to find. The answer: probably right at the centre of the 4D sphere. Presumably the more advanced a being becomes the further towards the centre it progresses.

The universal sphere may contain a sort of "psychic ether" that has a vibrational frequency increasing towards its centre and this ether responds to activity where thought equals action and like attracts like. The composition of the etheric matrix that fills the metaphysical universe is a big mystery. It is the vast underlying reality to our universe that we cannot see from here and is an area where most of our scientific investigators are unwilling or unable to go. However, some theoretical scientists like Ervin Laszlo (9) who refer to it as the Akashic Cosmos have attempted to do so and written on the subject. Once we are released from our physical bodies what is our existence like? My conclusions on this matter are drawn from the many anecdotal sources of evidence which are hard to substantiate without testable and repeatable data but nevertheless should be seriously considered.

The Soul
(Astral or Real Body)

I prefer to describe the soul as our real body until something better comes along, it is our life-force. The word soul carries a lot of religious baggage and may be used in new age mumbo-jumbo to describe something ethereal and ephemeral. Sometimes the term is dismissed with a joke or maybe associated with the sort of spiritual hijinks experienced during Victorian seances. Of course, to religious leaders and most theologians this concept of our existence is not in dispute, but modern science is not happy because proof is required. Preferably laboratory based and repeatable. However, there are a small but growing number of neurobiologists who are now considering the possibility that the mind is more than just the sum total of brain electrical activity and may exist beyond the brain as what theologians call the soul. If our bodies do have a soul, it is not easy to locate, unless we attempt to use psychic means: mediums, near-death and out of body experiences or hypnotic regression. There is in fact no shortage of space inside a human body, which is made of normal atomic matter, to place a metaphysical body. Apart from the freakish conditions in places like neutron stars the distances between atoms in everyday objects is phenomenal when compared with their actual size. For example, if the nucleus of an atom were the size of a golf ball, then its outermost electrons would be 2 miles away and of course the nucleus of the adjacent atom in the structure would be another 2 miles away according to Timothy Ferris (10). So, the vast majority of a human body is empty space. When two physical objects collide with each other what is happening in effect is that the electrostatic charges around one solid object are bumping into the electrostatic charges around the other and preventing further movement. Obviously, a soul is not made of atoms and molecules but whatever its composition there is no shortage of room for it. One enterprising American doctor, Duncan McDougall, 100 years ago weighed human bodies prior to and immediately after death and noted that they became on average 20 grams lighter, concluding that this was the weight of a soul (11). Even I am a little skeptical about that one.

Our souls are most likely comprised of an energy with a metaphysical vibration and while on Earth this metaphysical body is enclosed in a physical body for the duration of our lifespan. Interestingly our "deaths" in the physical plane will be the only instance during our eternal existence where an empty body is left behind on transition to a different level. Moving between astral levels, upwards hopefully, will be achieved with a change in the vibrational frequency of our metaphysical bodies. The new positions we reach will be at a point where we resonate at the same harmonic as one of the many zones surrounding our planet in 4D space. The "higher" this frequency the further we have travelled. To any observers present, when a soul moves from one astral level to another, they simply vanish and would reappear in their new surroundings and no corpse would be left behind.

Evidence For An Afterlife

As outlined above, information about life on the other side or afterlife is vague and sketchy. However, there is a huge amount of it. A visit to any large bookshop will reveal many shelves in the *Mind, Body and Spirit* section with many publications concerning the afterlife. The problem is what should be considered plausible and what is fantasy or deliberate hoaxes. Hard-boiled skeptics won't believe any of it and it is difficult to think of any evidence that would make them believe. When they do have a personal experience of psychic phenomena, they are often a bit stunned and still deny it and will invent even more implausible explanations to explain it away. There comes a point with so many personal accounts of psychic and spiritual experience that we have to say there really is something going on.

The phenomena of hauntings and poltergeists suggests that the non-physical entities who perpetrate these events are able to accomplish these manifestations that amaze or frighten us, because they have the benefit of an extra dimension to move in. There have been many well-documented cases of poltergeists causing objects to appear and disappear or apparently "throwing" small objects eg. pebbles or coins. Frequently the thrown object is not seen in trajectory but rather "appears" at its target which suggests that it does not use our time and space to get there. Imagine an entity in the afterlife collecting a 3D pebble with a 4D hand, what would we observe? Well, the pebble would vanish from our locality and reappear wherever the entity wished. In fact, the pebble is withdrawn from 3D space via the

4th dimension and is then replaced but would disappear during the process. This pattern of events is typical of poltergeist phenomenon. It also appears that some energy is required from the physical world to achieve this action as the objects in question are normally at considerably different temperatures from the ambient environment. Such an explanation also raises the possibility that in principle much larger objects or even people could be transported in this way. A parallel situation was explored by Edwin Abbott who illustrated what people in a 2-dimensional existence would observe if a 3- dimensional object interacted with their universe in his book *Flatland*.

When mediums give readings for people who have lost relatives there is a fairly common question directed at those who have gone to the afterlife or "passed over". That is: What do you do with your time in the next place? The answers typically include the explanation that from where they are now, there seems to be no time, or it is perceived very differently, certainly passing at a different rate; things don't change much. This difference in experience maybe explained by the change in perspective from existing within a 4D volume to that of being on its 3D surface. The 4D environment with its extra dimension expands as a volume not as a surface and this may explain the altered perception of time in the afterlife when compared with the physical universe.

There is further anecdotal evidence of this alteration in time perception from the many reports of hauntings in some specific locations where trapped spirits repeat their daily rituals sometimes over many centuries and even longer. From the point of view of our physical world this behaviour seems rather tragic. But for them, time may pass at a considerably slower rate. Some useful reading here is *Afterlife Adventures* by W. F. Martin (12).

Out of Body Experience

A small proportion of the population has out of body experiences (OoB) and a fewer number can do this at will. Similar phenomena are also reported by some people who have near death experiences (NDE). This phenomenon was first described by the American doctor, Raymond Moody (13). It usually occurred during a medical crisis where the patient was often pronounced clinically dead at some point during their NDE. Dr Pim van Lommel, a now retired cardiac doctor, has extensively researched this subject among patients who had suffered a cardiac arrest and has published some very interesting findings in the medical literature and has done interviews which can be found on YouTube. The NDE differs from OoB in that it appears to be largely involuntary, while those practicing OoB travel can choose to some extent where they wish to go and what they want to do. Investigators from the Monroe Institute, in particular their founder Robert Monroe (14) have produced some of the most plausible locations for the destination of souls in the short to medium time interval after physical death. Their OoB explorations have led them to conclude that planet Earth is surrounded by a series of zones extending outwards. The outermost regions are inhabited by those who are more spiritually advanced and would have few if any further incarnations into the physical world. Proceeding inwards these regions are occupied by people who are still attached to the experience and desires of physical life and may exist here for long periods before returning to another physical incarnation. The innermost regions contain what Monroe described as The

Belief System Territories where many religious people who are infatuated with scripture or fundamentalism have become trapped. Further in are "The Hells" which are self-imposed. The zone immediately adjacent to Earth is occupied by what we term ghosts, many of whom do not realise that they have died. The frequently recounted tunnel of light from those who have near death experiences is a short cut that takes those who have left their physical body directly to their higher sphere. If you do not take this route, you may remain earthbound and this occurs to a small percentage of people.

I would maintain here that these regions are spheres surrounding our planet *but not* in normal physical space and time. Also, information that comes through from people working as mediums report that the sun, moon and stars are not visible in the afterlife, which should be expected in this situation, unless you wish to visit them at close quarters. For example, travelling, say halfway to the moon in a spacecraft and looking for these zones will not work.

The tunnel of light experience that people have on death or during a near death event is another indication that there is another direction of travel available. Monroe felt that he was somehow becoming out of phase with his physical body during OoB travel but he also found that he could travel in a direction not describable in physical space/time. Monroe also noticed when travelling to the Earth's more distant astral spheres, that when he looked back, he could see small blobs of light leaving and returning to the planet. These were the non-physical components of individuals after physical death and those beginning a new incarnation. There is also a lateral direction and distance available in the metaphysical dimension through the compressed and overlapping vibrational spheres leading to specific zones, a finding that has been put forward by others in *The Joseph Communications* (15).

These inhabited spheres are located in 4-dimensional space/time and require a direction of travel we do not have access to with a physical body. Travelling outwards from the Earth in all directions through the spheres would eventually lead to the centre of our 4D universe, although this distance would be enormous. The spheres surrounding our planet in 4D space are represented in the diagram below. Obviously, it is difficult to show 4D space on a 2D page.

Another investigator, Michael Newton in his book *Journey of Souls* (16) used hypnotic regression to reach the memories of subjects' previous lives and their existence *between* physical reincarnations. One of the repeated assertions of the subjects under deep hypnosis was that their location in the afterlife was in some way bounded by a region of curvature when they traveled long distances which could be interpreted as being the internal boundaries of a 4D sphere surrounding our planet. The prolific author Dolores Cannon (17) has also used hypnotic regression to investigate the afterlife and has published many books and produced YouTube videos on this subject.

Monroe described some of the innumerable different regions inside the higher spheres. The landscape and surroundings appeared solid but were clearly not made from physical matter and these could be altered in some way by focused human thought. So, we would find anything that could be imagined here on Earth and this was continuously unfolding with time as humanity progressed. In fact, it is so comfortable that many stay there for long periods. Mountains, lakes, vegetation, beautiful houses, cars, whatever you wish. These mentally constructed environments that we inhabit are not affected by our time but do require some continual mental input by their creators to maintain them. There are problems in some of the inner spheres depending on your level of spiritual awareness and burden of guilt and your situation is determined by your preoccupations at a mental level. In such regions your freedom of movement is restricted and you will be in the company of similar people. In the extreme cases these are described as the Hells by Robert Monroe (18). However, existence in the inner spheres in not necessarily forever. People from the higher spheres are constantly attempting to retrieve those from lower down which is frequently difficult because of the metaphysical vibrational differences. Sometimes those trapped do come around to their situation and make attempts to leave, whereupon they can be rescued. Such individuals are likely to choose further lives on Earth in order to progress and may incur what is sometimes called karma, or more simply put, a balance of experience. Spiritual guidance from much "higher" sources is often available during this process. Beyond the Earth spheres is the ultimate destination for us all.

Figure 2. Afterlife Spheres Around The Earth

How The Metaphysical Dimensions Work

The Afterlife dimensions are definitely not static environments; we do not sit on clouds playing a harp or coast into some heavenly realm, after what we consider a hard life. In the lower metaphysical levels, there are the immediate astral regions and as the metaphysical vibration increases, they transform into higher astral, celestial and heavenly dimensions where we become much closer to our Creator. It may be sub-divided in other ways, for example laterally, to include different parts of our cosmos, surrounding other planets, but these places are all *non-physical*, ie. there are no atoms, molecules, gravity or electromagnetic radiation as we would understand in the physical universe. An equivalent to gravity is a particular problem as it seems to exist in the Afterlife but cannot be related to objects with mass-warping space-time in the physical universe as we experience it on Earth. However, the higher a person travels upwards, through the astral levels, the less they become affected by the metaphysical equivalent of gravity. Exactly how all the conditions in the metaphysical dimensions mimic those that exist in the physical universe is to us at present unclear. Our lives in the metaphysical seem to continue in a way that is not too dissimilar from the situation we find ourselves in here, except many of the rules are different. Especially in that our environment is reflected by individual metaphysical vibration which is determined by what we think and what we have done previously. The worlds of the afterlife dimensions consist of a form of psychic (spiritual) ether that becomes more refined as we progress towards the centre of metaphysical creation. These metaphysical afterlife dimensions are perfectly organized with a divinely inspired order, hierarchy and rules which will surprise most of us when we return. But I am happy to predict none of the depressing bureaucracy that plagues our lives down here. The commonly used expression "as above so below" was originally attributed to philosophers in the ancient Hermetic Texts and refers to that which manifests in the higher metaphysical realms will then transfer down to Earth. Strictly speaking this is the wrong way round. It should be: "as below so above". It is the basic structures, created here on the Earth plane by us and the natural forces of planetary evolution, that carry over to the metaphysical levels to give a similar form and function but enhanced at higher astral levels. So, in a way things start here and that gives us a special responsibility. It could not be any other way as everything we see around us here in the physical plane has taken a vast amount of time, billions of years, to come about by natural evolution and evidence of this process is clearly visible including human development. It was not imposed from "above".

A question often asked is what are things made of in the non- physical afterlife as obviously there are no rocks, soil, metal, wood or water in these realms. The collective wisdom obtained mainly from a number of mediums is that, from the metaphysical ether something referred to as the "essence" can be extracted, which is mentally malleable and can be used for construction. The people there can acquire an expertise with this material similar to what we have here and are able to provide others with artwork or objects unique to their ability. In some planes even food and drink are produced, however probably for nostalgic reasons, as we don't actually need it.

The importance of our experience in the physical worlds of existence is that our reality is created through thought, followed by effort and work, with knowledge gained in the process. Often described as the burden of life in our human condition by Jesus in the Bible. However, in the metaphysical dimensions thought equals reality immediately, which can be heavenly or hellish according to our astral/metaphysical resonance and we are the architects of our personal situation there. We may not fully understand why, but our Creator decided it was necessary, that is the way God made it; an explanation not popular with atheists.

When we die on Earth in the physical dimension, we leave a body behind; a unique event in our total existence and will of course occur at the end of each reincarnation. Obviously in the metaphysical life no bodies are left behind because moving from one astral level to another involves only changes in vibration or frequency of our metaphysical bodies. Curious people often ask what earthly activities are still available to you in the metaphysical afterlife. Can you die, have sex, have children, play sport? So obviously you cannot die. In the lower regions where unsavoury individuals are found, they will tend to continue their old criminal or depraved lifestyles that they had while on Earth, unless they seriously wish to change. Everyone there will think the same, so murder, violence and extreme persuasion as part of their life pattern may still be attempted. But of course, brute force cannot work if there is no real physical body to overpower. Anyone "killed" immediately reappears. How the people inhabiting these places deal with this I do not know. There are well described haunted earthly locations where old battles continue to be fought, for example near Gettysburg, Pennsylvania, where discarnate soldiers still seem to be fighting the American Civil War. Spectral Lancaster bombers from World War 2 have been often reported from the Derbyshire area of England. There are several locations in England (eg. around York) where ghostly Roman soldiers occasionally can be seen still marching, interestingly on their knees. They appear on their knees because the ground on which they originally marched is now at a higher level and they still perceive their old path from when they had physical form. But why are these old battles still being fought in some lower astral levels, especially around 2000 years later? The obvious absurdity of fighting to death with bodies that cannot be killed seems to be overwhelmed by the emotional attachments made during the repetitive rituals and group bonding that occurred during their time spent in the armed forces while on Earth. But using force to prevail and impose your demands and views on others, cannot succeed in the afterlife environment. Some paranormal observers believe such phenomenon are merely astral (Akashic) recordings of extreme events attached to a particular location. But other psychic researchers claim that these ghostly apparitions sometimes have an awareness and can react to people in the physical plane who are watching them, so what is witnessed is not a simple recording. Who knows, maybe these repeated battles are what those who have passed on regard as a fun day out. Re-enacting old battles is sometimes done here too by various theatrical groups.

There is certainly an absence of politics in the astral realms which will be a relief to most of us. For the simple reason that resources are infinite. If you want something all you have to do is think of it and no regulation or limits are required, provided that you have arrived at a reasonably progressed vibrational level. If you wish, you can have a mansion by the sea with a tennis court and a Ferrari, but of course so can everyone else, so it is not going to feel quite the same. However, according to OoB travelers in the lower astral realms the quality of dwellings and the environment is much poorer and is a reflection of your metaphysical frame of mind. But change is possible. There is also a much greater freedom of thought, although communication there is more like spoken telepathy. Individuals who have had OoB experiences taking them to the higher astral realms report that the people who exist there have become in a way "mentally transparent." There are no secrets there, we cannot lie and our thoughts and feelings are automatically exposed to everyone we communicate with. Presumably in this more celestial environment the concept of confidentiality and privacy is not important or relevant. What it also revealed to those we interact with is the record of our many previous lives, so nothing is hidden, good or bad. I suspect most of us here on Earth would find this rather uncomfortable. Our incarnation here in the physical realm will have at times made all of us regret our actions and there is really no escape from this. Maybe there are ways of keeping thoughts private but I suspect the more advanced astral levels are only occupied by open people who are happy with things being that way.

So, individuals who have similar ideals and backgrounds are going to resonate and occupy the same astral level; like attracts like. Politics can be very divisive on Earth particularly between "left" and "right." But our position in the astral realms will reflect our true inner nature, rather than how we have voted on Earth. What we have done with our wealth, or service to the overall good of humanity, will automatically count for much in our afterlife destination. Jesus would certainly have been a socialist 2000 years ago but today? Maybe not; a tricky question.

One of the purposes of reincarnation is to bring people together who have been living in rather separated higher and lower astral spheres. Because they are on very different vibrational levels, they would normally be unable to encounter each other. However, such people can interact again if they both reincarnate in the physical plane. The physical dimension is a great leveler and forces certain people to come together again, particularly if there are karmic issues which have to be resolved.

Sex can certainly be continued in the lower astral regions. Our earthly physical bodies are still as we imagine them so can also function in a comparable way and maybe the experience is heightened. Jurgen Ziewe (19) described an encounter during one of his OoB travels where a sexual offer was made to him by a lady inhabitant he encountered in one of these regions, but he declined. It is not possible to have children as we would understand it, as children are born through the process of reincarnation into a physical body. But according to the readings from Michael Reccia of the *Joseph Communications* there are sometimes situations where traumatized parents arrive on the astral planes who have lost children and are given a child temporarily, to help them recover. These could be babies who died very young or other souls who volunteer to be children for a while.

As for sport I do not know. We would not tire there as we do here and sleep is not necessary. Most sports do require a reasonable amount of the Earth's gravity, but the mental forces emanating from participants and spectators would have a profound effect on the movement of a ball so physical skill would have little to do with the result. Certainly, a form of gravity or the astral equivalent of gravity is somehow recreated in the metaphysical dimensions. Gravity in the physical universe, as described by Einstein, results from a large mass warping the fabric of space-time, creating a form of "well" or depression into which less massive objects will fall. OoB travelers like Jurgen Ziewe observe that the lower astral regions do have what we might call normal gravity. In fact, he noticed that normal activities like walking or climbing stairs in the lower astral regions required a comparable effort to what we might experience on the physical plane. The residents in these regions existed as though they were on Earth. In some cases, travelling by car was preferred to walking even though a car is just a metaphysical mental creation. In the higher realms although a form of gravity exists, we somehow become "lighter" and flying is possible. How gravity is created when we leave a physical universe without mass and energy, we can only speculate on but must be a product of the vibrational mental consensus produced by inhabitants in that region. The translation of metaphysical abilities from the astral dimension to the physical dimension is often observed during encounters with the more developed ETs. In particular, telepathy and "floating" above ground level. This is likely to reflect their prolonged periods of OoB experience which somehow is carried over to the physical dimension; something hardly any of us here now on Earth is capable of doing.

People with OoB experiences describe how in the metaphysical dimension, objects and their environment could be manipulated or altered by focusing their attention on them. For example, a car could be visualised and it would appear, and if you opened its bonnet the engine would be there, *just as you imagined it would look*. So, you find what you expect to see from your memory or the collective memory of the inhabitants in that plane, for example the buildings there are often created by a group of people who exist in that location. Jurgen Ziewe on his travels around some of the inhabited metaphysical planes noticed he could actually alter some small parts of buildings he examined, but sooner or later the structure would return to its original shape. The concept of altering your surroundings by the power of thought does not work in the physical universe. Possibly for good reason: that is why we are here, the physical universe, like where this book exists, provides a unique environment for individual exploration away from the Creator's realm. However, on the tiny quantum scale it may happen, eg there is the wave/particle duality of an electron or photon. The state of the photon is ambiguous as either a wave or particle until it is observed or measured. The process of observation seems to crystallize the photon into either possible state once it is observed, by a human anyway (the observer effect). Interestingly this phenomenon also occurs when the observation is made by an independent recording device and in the process the universe acquires information. Who knows, more advanced minds may be able to extend this power of mind over matter to modify more substantial objects in the physical world. There is some feedback from people who are able to achieve extended OoB experience that they begin to acquire other "paranormal" abilities like those observed in advanced ETs. There are many good reads for anyone wanting details of how life carries on in the metaphysical realm. Paul Beard in his book *Living On* (20) refers to our surroundings in the metaphysical dimensions as being somehow malleable or *ideoplastic* when mentally focused upon. This can be heavenly or hellish for us depending on our level of personal advancement with mental or karmic baggage. Further very plausible descriptions can be found in *Afterlife Adventures* by William Fergus Martin also *Resonant Mind* and *Life Review in the Near-Death Experience* by David Lorimer (21). As ever, with the descriptions offered in such books, the reliability of the mediums who provide the information is crucial.

An afterthought to this chapter is a word of what might be useful advice to us all. Inevitably the day will come for each of us to make that transition to the next existence. On leaving a physical body there is a risk that some individuals can become "stuck" in the lower astral regions and become what we call a ghost. This situation is self-imposed, possibly due to feelings of guilt, sometimes a strong earthly attachment with a reluctance to move on, or it may be a genuine ignorance of what has happened to them. Typically, passed on friends and family will be there to guide us but sometimes they are ignored. However, there is always an escape route to those who wish to move on but are not sure where to go. Look upwards and somewhere there will be visible a small bright star or speck of light, when focused on, this will expand to the light tunnel that will pull you up to your natural destination.

Our Metaphysical Origin

All life was created by the Source (God) but exactly how, is still rather mysterious, with particular reference to ourselves. There are two main possibilities.

The first is that the Source or our Creator fragmented some, or all of its Being into trillions of buds and scattered them throughout creation; each of these buds being an intact human soul. Each of these souls was given sufficient intelligence and freewill to exist in the physical universe for a limited period of time before returning to the metaphysical dimension to reassess its progress and carry on. Presumably this system is continuing as more humanoids are born throughout the universe. Such a process would be expected to include other "lower" life forms like animals and presumably all the way down to insects and even unicellular life. All life seems to have a metaphysical component (the word soul in this context may not fit well as it is only applied to humans). Visitors to the astral realms who have made OoB journeys report an abundance of animal life forms. There are many books on the afterlife that discuss the destination of animal life forms, especially pets. An amusing account came from Jurgen Ziewe who arrived during one of his travels in a lower astral location where some cows had also ended up. The locals considered building an abattoir to process them into beef, but fortunately decided against it, thinking they must have already suffered enough.

The second possibility is that the advanced human soul has accumulated gradually from smaller, most likely animal souls, throughout evolution. These may coalesce together after many reincarnations to become candidates for the more sophisticated and complex process of human reincarnation at some point. The implications of this pathway would be very significant. For example, at what point in our evolution did pre-Homo Sapien primates become defined as human, therefore becoming acceptable vehicles for human reincarnation? Maybe in the time of Australopithecus around 4 million years ago? The spiritual self- assessment and self-judgement process that we all go through after physical death can hardly apply to animals which have their behaviour determined by instincts rather than by considered moral choices. Do higher animals have a life review and how far down through the complexity of life would this apply? So, the switch point from animal to human must be fairly well-defined, maybe when our brains had reached a certain size. This must have started a few million years ago on Earth when we were say, Homo Habilus, H. Erectus or H. Heidlcbergensis. Also, if animals eventually progress to humans, then our treatment of them would need to radically change. Should we therefore regard all animal life as potentially becoming human at some point? Is eating meat cannibalism and should we become vegetarian? I think this is a rather complex theological subject beyond the scope of this book.

The Lower Metaphysical Region

Some OoB travelers have described certain lower astral planes as gloomy, desolate and inhabited by rather unsavory beings. Along with the entities that are by-products of negative human thought and emotion, are the dark human souls. By contrast, in the higher astral realms, the region or level you reach is determined by what has preoccupied you in your physical life and what reflects your thoughts and mental frame of mind. It can be a place of harmony, beauty and peace as described by Lynda Cramer, *Five Years in Heaven* (22). However, inhabiting the lower astral places are those people who have deliberately shunned the spiritual light along with darker entities, who are not original physically reincarnating beings created from the Source. The darker non-incarnating entities occasionally show up in physical locations and maybe associated with hauntings or poltergeist-like phenomena. We often label these entities as demons or dark elementals which have managed to take on an independent existence. The origins of these entities are from people with negative or hateful emotions projecting their thoughts and feelings into independent manifestations, which develop a lower astral life of their own. The people who inadvertently produced such astral forms will remain pestered by their creations until they can be dissolved. Similar entities are described in the Buddhist religion as Tulpas which are thought forms created in the minds of humans and eventually acquire a life of their own. However, there are also some people who deliberately attempt to create dark astral entities for their own purpose such as the famous occult practitioner, Aleister

Crowley. A typical appearance of a dark entity is the sighting of a jet-black shape, which maybe humanoid, often with a fuzzy outline. It looks so incredibly black that light is totally absorbed and none is reflected, like a black hole. These entities have more recently been associated with many haunting phenomena and have been called the Shadow-People. These non-physical entities are probably showing up more frequently due to the widespread use of camera phones and CCTV. They have a poltergeist-like behaviour and are seen around human dwellings. They are often noticed fleetingly out of the corner of the eye, where there are human observers and their favourite prank is peeping around corners. Our holy teachers would correctly denounce these manifestations as something not-of-God. Hopefully when the new spiritual light illuminates this planet, these shadowy regions and their occupants will be eliminated and swept clean. Of course, there are many dark human souls in the lower astral regions trapped in the "Hells" as they were referred to by Robert Monroe. With the new light they will be forced to confront their negative choices and move on. Exactly what fate befalls the worst individuals of humanity is not clear but in principle every soul is saved.

There are of course many lower regions occupied by non-physical human entities that we call ghosts, who are unable to move on. It is common to find certain locations on Earth, like buildings or battlefield sites that seem to have more than one ghost present. These ghosts may not be aware of the other discarnate beings in the same location, especially if they are individuals who have died during different eras. People existing in the lower astral regions can have different shades of metaphysical vibration so they will have no interaction with each other. For example, an ancient battlefield site in Britain may have the ghosts of Roman soldiers appearing but later this place became a World War 2 airfield, where deceased airmen also appear. However, the ghostly Roman soldiers and the airmen would not interact, or be aware of each other, which we may regard as a bit strange. A similar situation may occur in old buildings with multiple ghostly occupants, who may have got stuck there over the years. Very often these ghosts seem unaware of each other although they can see people with physical bodies.

Life Forms in the Lower Metaphysical Regions

Apart from the discarnate human inhabitants there is a veritable zoo of other creatures existing there. All the byproducts of human imagination that have taken form over the ages can have an existence there. For example: fairies, pixies, elves and the equivalent from all the Earth's cultures would end up there. However, these lower astral zones can partially overlap with our current physical plane and we can, at times tune-in to the "other side". As well as us physical inhabitants tuning into the metaphysical dimension the process may be two-way and the lower astral somehow intrudes into the physical dimension, so these creatures seem just as surprised as we are when they are spotted. It is not really clear why this happens and it is often associated with particular earthly geographical locations. For example, nodal points on Ley Lines. These creatures should not be confused with animal ghosts. There are numerous cases of pet owners seeing their pets some time after they have died, like showing-up in an outdoor family snapshot or sniffing around the house. These pets seem aware that they are no longer existing in the physical plane but nevertheless like to turn up for old times' sake in their old home but only as a metaphysical presence. Domesticated animals may feel encouraged by the familiar human surroundings and their metaphysical vibration would resonate with their previous human owners and make it easier for them to appear. There has been some research into this phenomenon by the British Psychological Society and a number of books eg. *Pet Ghosts* by Joshua Warren (23). Probably wild animals could do this as well and this may explain the sightings of unusual animals that seem to vanish. For example, the Loch Ness monster could be the ghost of a plesiosaur and similarly the recent bizarre footage of a flying pterosaur caught on camera. See TV series like *The Paranormal Caught on Camera* for many of these peculiar sightings.

Cryptids

The wider definition of cryptids includes bizarre zoological animal types that may not be part of the Earth's normal fauna and have a supernatural origin. So, when there is an astral/physical overlap in places, this can enable the appearance of some inexplicable animal types or cryptids. There are many reports of baffling cryptids that are seen but can never be captured or killed. These are quite well known and include the yeti types and the numerous moth-man sightings in West Virginia, North America, where they even have a dedicated statue. Reports of yeti type primates are springing up all over the planet. The first modern reports were from the high altitudes of the Himalayas but they have now been seen regularly in the Pacific Northwest of the US where this unusual primate is referred to as Sasquatch or Bigfoot. The Australians have one called the Yowie which is part of Aboriginal folklore and was initially described as an Australian indigenous ape but never officially documented. In Russia they have the bad-tempered Almasty which has been reported in the unpopulated regions for at least 300 years. One possibility is the Yeti groups scattered around the globe could be the remaining populations of a giant primate known as Gigantopithecus that allegedly became extinct around half a million years ago. However, there are problems with this hypothesis because the many sightings of yetis do not match the Gigantopithecus phenotype made from reconstructed fossil remains. Obtaining hard physical evidence of these creatures has been extraordinarily difficult. In particular the encounters with Bigfoot in North America.

Mountain walkers and hunters report the strong odour and noises these creatures make so it seems they can have a very real physical presence when they wish. But on the few occasions when they have been shot at, the bullet apparently has no effect, passing straight through or the creatures simply vanish. This suggests there may be a paranormal quality about their existence and they can revert to a non-physical dimension at will. Paranormal entities probably can only be stopped with some sort of fantasy weapon like that used in the film *Ghostbusters*. Maybe it is possible to construct such a device, if only we had the opportunity to test it. The appearance of these odd creatures seems to be increasing but that may be due to the more widespread presence of cameras. Possibly the veil of separation between the physical and metaphysical realms could be breaking down or even being slowly lifted. Sometimes there seems to be a sort of "fuzziness" of the lower astral dimension overlapping with the physical universe.

UFOs are frequently associated with the appearance of Bigfoot and other unusual cryptids capable of exhibiting different forms of paranormal behaviour. This is unlikely to be a coincidence, but we do not know as yet, how the two manifestations are connected. One possibility which I would like to suggest is the secondary effect of interdimensional travel or the space/time distortion technology that UFOs employ for propulsion or instantaneous appearance/disappearance. This extreme distortion of space-time may also "drag" the surrounding lower metaphysical dimension along with it causing a dimensional overlap. This intense field of distortion produces a form of interdimensional gateway, opening a portal that enables lower metaphysical life forms to come through and appear in the physical world on a transient basis. This might be a deliberate action instigated by ETs or a by-product of their proximity. There are instances of people who claim that when abducted by the Grey ETs they were somehow transported through solid objects like doors or windows during the abduction process. We do not know whether the ETs used some advanced device to facilitate this process or whether they employed a psychic manipulation of solid matter to render it permeable. But as mentioned before, solid matter is actually far from solid when examined at an atomic level and there may be a way to exploit this, so permitting physical objects to pass through each other.

The US government and other scientific authorities have shown a spectacular lack of interest in investigating the authenticity of Bigfoot's existence and this does look rather suspicious, especially in the light of thousands of reports; maybe they already know. Allegedly some US security agencies have held in secret locations, the preserved corpses of a few Bigfoot individuals that were collected after the Mount St Helens eruption in the state of Washington in 1980. You might just think that in the interest of contributing towards one of the most amazing discoveries of the century these bodies would be handed over publicly to scientific experts in comparative zoology for some detailed studies. Interestingly not; so why not? Very possibly when examined genetically and anatomically the scientific experts may well come to the conclusion that some of these creatures could not have originated on Earth. There were many close encounters with Bigfoot-type creatures in the Pennsylvania area of the USA during the 1970s. The Bigfoot footprints observed in the US are normally similar to humans but much larger. However, the footprints found with Bigfoot associated with other paranormal phenomenon have only 3 toes, suggesting they originated on another planet and the investigator Stan Gordon has published widely about this (24). The witnesses in Pennsylvania frequently reported UFO activity in the nearby area around the time of their encounters. Furthermore, when some of the local witnesses described their Bigfoot encounters in detail, with the public, they then experienced Men- In-Black type visitors who seemed rather intimidating and told them to stay quiet about their encounter.

The reasons as to why ETs want to let a few Bigfoots loose around our planet does not have a snappy answer. But challenging us and even winding humans up a bit, does often seem to be part of their overall plan. My personal theory regarding Bigfoot and UFOs maybe a bit outlandish but pertains to primate physical and spiritual evolution. On this planet there is little in the evolutionary primate hierarchy between humans and the higher primates eg. chimpanzees or gorillas where our brains are 4 times larger. The Bigfoot species is going to fill that gap and could be regarded as a final step in animal evolution before fully intelligent humanoids are reached. The Bigfoot seem quite intelligent, are allegedly empathetic and have strong family instincts but they are never going to understand differential calculus, compose sonnets or comprehend the finer points of civilization and a sophisticated society. ETs fully understands the workings of animal reincarnation and is attempting to develop the process further on Earth.

If there were a well-documented event where a Bigfoot badly injured or even killed a human in the wild, then a massive backlash with a concerted and thorough investigation is likely to follow. Of course, if this involved the press, police and professional search parties it would be the last thing that a Bigfoot population would want to happen and maybe they know this, which explains their extreme reticence and preference for avoiding any human interaction. The situation with the more recently observed larger cryptids, as mentioned above, is that they leave a much firmer imprint in the physical world. They seem to have a supernatural origin yet are not entirely composed of flesh and blood. For example, there are numerous accounts of cars being chased by the Mothman and some have described it landing on the vehicle and doing physical damage to the metalwork and paint. The Bigfoot or Sasquatch can break trees and leave footprints in snow and water cryptids clearly leave a wake in the water they inhabit. Mothman and the yeti types are presumably in principle at least partially carnivorous, and mysteriously mutilated or eaten bodies of different animals are sometimes found in those areas that such cryptids inhabit. It is a puzzling question as to why apparently semi-physical cryptids need a totally physical food source. If a Himalayan yeti can kill and eat a yak, why can it not be stopped by a bullet? The Mothman and another cryptid called "Goatman" reported on many occasions from several states in the US allegedly have glowing red eyes. There are no known earthly animals, or humanoids for that matter, which have self-illuminating eyes. Eyes maybe very reflective like in cats but they cannot generate their own source of electromagnetic energy. The Jersey Devil is another cryptid seen on numerous occasions in the Pine Barrens of the US state of New Jersey. Witnesses have described its appearance as peculiar, rather like a "3-dimensional shadow". So, this and similar creatures could be completely metaphysical entities rather than transiently composed of flesh and blood and are unlikely to be part of our physical reality. Many of them have been witnessed by locals for long periods of time, maybe over hundreds of years. If this were the case there must be a breeding population which seems rather unlikely, as this would have been noticed by now. Like with Bigfoot, some of these cryptids are seen with UFOs spotted in the same vicinity so there may be a connection between them.

Probably ETs like to test us occasionally with conundrums of high strangeness to see how we react, or maybe for their own amusement. A good example of this type of situation was the goings-on at Skinwalker Ranch, Utah, in the US which left a team of well-qualified scientific investigators baffled. With smart phones and CCTV cameras now being so widespread there has resulted some fascinating clips of such unusual creatures being caught on camera and hopefully these will increase and may reveal further clues as to what is going on. Recorded video footage of this sort of phenomena can be seen on TV channels like *The Paranormal Caught on Camera*. Also, many YouTube clips should be checked. Some will be hoaxes but most seem credible and are currently inexplicable. Many of the odd cryptids picked up on CCTVs and mobile phone cameras are not spotted until a later time when the movie footage is reviewed. This suggests that they may not always be visible directly to the naked eye. There is a noteworthy and inexplicable example of this from Indonesia available on the *Blaze* Paranormal UKTV channel. A rather grotesque humanoid can be viewed clearly, where it also produces a clear shadow, on the screen of a digital hand-held camera but was completely invisible when looked at directly by all the people present. This being was supposedly a legendary humanoid cryptid known locally as the Tuyul. Why a digital imaging system can detect such a creature but the human eye could not, is implausible. It might be of further interest to paranormal researchers when investigating cryptids to use a *Google-Glass* type of technology with a digital interface that has alongside, a real time visual perception, so that both images can be viewed simultaneously.

The author and paranormal researcher Anthony Peake investigated some of the many life forms that inhabit the lower astral regions and has called them egregorials in his many publications (25). He believes that the creatures in this region are not completely independent astral life forms but are somehow linked to humans who have manifested them. Other people experimenting with OoB excursions and remote viewing have also encountered these entities and refer to them as *paraphysicals*. Peake observed that particular lower realms of the astral dimension can be experienced by people when under the influence of DMT (dimethyltryptamine) commonly referred to as ayahuasca, which is a hallucinogenic drug used by some native tribes in South America. The drug allegedly induces the NDE and OoB experiences in studies on volunteers. However, the paranormal experiences induced by DMT did not seem to be taking the subjects to a metaphysical level that was particularly helpful in terms of spiritual enlightenment, understanding and progress. They were stuck in the presence of often unpleasant low-level astral non-incarnating entities.

The Occult

The occult refers to the study and practice of certain darker aspects within the lower paranormal realm. Occult means hidden and it covers a wide range of practices like real magic, witchcraft and devil worship. It could be called the "dark side" of the paranormal and often sounds vaguely sinister, especially as practiced by the infamous occultist, Aleister Crowley. However, in reality it means those attempting to practice the occult are simply tuning into forces of the very low astral realms. But in this region, most of the inhabitants are not highly evolved spiritual beings but a kind of metaphysical residue from the rest of Creation. Sometimes those people who have had near death experiences have found themselves in an ascending tunnel of light but when they looked around on the way, they could see some of the lower astral regions and the life forms present beneath them. These life forms are created as a metaphysical by-product of the human mind and occasionally they can be seen in the physical realm. Those who indulge in occult practices are deliberately attempting to force the metaphysical realm to overlap slightly with our world. Any creatures that manifest seem to have some measure of intelligence but are all essentially by-products of humanity's mental projections throughout the ages. Some of these discarnate beings are quite disagreeable and would include creatures like demons and night crawlers which have been caught on domestic CCTV cameras, especially in the USA. They are unable to incarnate into physical bodies and will never reach higher astral realms.

An interesting example of people manifesting a non-physical entity was the "Philip" Experiment. In 1972 where a Canadian paranormal researcher, Dr. G. Owen conducted a study to test his theory that, "Ghosts have an objective reality, but they are created out of the minds of those who see them". The experiment was so successful that parapsychology researchers went on to repeat it and generate other metaphysical entities with different groups of investigators. It also was the basis for the 2014 film *The Quiet Ones*.

Unfortunately, there are many people who wish to get involved with the occult and Satan worship but it is unlikely to do them much good. One of the problems which casual investigators have, when attempting to communicate with other side, is that they assume everyone there has complete access to all information. We may hope to receive life changing revelations from a wise and ascended beings of light or information about deceased loved ones.

This is a serious miscalculation. But the occupants of lower astral regions do have a type of temporary advantage over us because they can see so much in the non-physical world that is obscured from us. However, their general knowledge is typically minimal and their ability to travel much restricted. Many people who have passed over are just as ignorant and perverse there, as they were during their physical life, especially if they have remained trapped in lower spheres. So, attempting to ask meaningful questions, eg. by dabbling with Ouija boards, planchette devices or an inverted glass will result in you receiving very probably worthless or malicious information. So don't bother. The entities who come through in such circumstances are often pretty dim and have little access to any relevant information and in fact some of them can be quite pernicious. We think because they exist in the metaphysical dimension, they must enjoy a panoramic view of the afterlife. This is far from the case, as they are not capable of visiting the higher astral realms of which we are curious and hopefully where our futures will be. However, some genuine mediums do seem capable of making a convincing connection with people who have passed over to higher realms and receive real information; but often what comes through is typically rather mundane, disappointing and trivial.

Pest Control

Hauntings and poltergeist activity can be a serious problem for the occupants of domestic dwellings where these phenomena occur. Not surprisingly, many people experiencing such events find this very distressing even terrifying and choose to leave their homes as soon as possible, I wouldn't blame them, but some others come to terms with it. The so-called "shadow-people" are a common occurrence and seem particularly unpleasant because they have a transient visible presence at certain locations within a home. Anyone watching TV footage from programmes like *The Paranormal Caught on Camera* will spot the shadow-people cropping up on a regular basis in domestic environments that have a haunting problem. They are almost black, rather short, have reflecting eyes and seem to enjoy peeping around corners at those observing them. These entities presumably occupy certain regions of the lower astral region and somehow latch on to particular locations or individuals and enjoy bothering them. I would suggest they are not apparitions of physically deceased humans and are unlikely to be capable of incarnating into human form. No one should have to tolerate these intrusions and paraphysical entities should not be allowed to petrify people living in their own home.

There should be robust and even aggressive methods available to banish them from human spaces. When the occupants of these infested homes try to remove the entities, they normally bring in paranormal investigators with a medium or even a priest. Blessing the house by reading from the Bible and sprinkling holy water may work if you are lucky in some cases. But if the shadow-people are not of human origin, then getting a medium in to ask them to move on to a higher place, is unlikely to work. What we need is some type of technology, maybe like what we had in the film *Ghostbusters*, but as yet, we do not know of any physical devices that affect or interact with the lower astral dimensions. I would suggest to anyone with a regularly occurring haunting problem to try some home-spun techniques in case they work. There are a number of possibilities. Firstly, there might be some types of radiation that irritates them like UV lighting or infra-red. Lamps that produce light in these regions of the spectrum are normally available in larger electrical shops. So, placing this type of illumination in those parts of the house where the haunting is mainly located may drive the entity away. Also directing a more highly powered laser pointer at the entity may scare it off, if you get the chance. Mobile phone jammers that produce a blast of multifrequency radio waves, which silence mobile phone signals, may cause distress to astral beings in the vicinity. Secondly, high frequency sound generators may work. These produce an acoustic signal that humans cannot hear but do repel small animals like mice. Thirdly, a room filled with smoke produced by burning incense or josticks might be worth trying if the haunting is located to a small room or zone in the home. Finally, a new device has been developed which is a hand-held electromagnetic pulse generator, that is almost *Ghostbusters* technology. It will be introduced shortly in the UK to stop e-scooters in their tracks when the riders are using them illegally. Such a pulse directed at an e- scooter motor confuses the control system and causes it to shut down. There is much anecdotal evidence that paraphysical beings and ghosts can interfere with electrical systems. For example, in some haunted locations domestic TVs, radios and phones have been witnessed operating, even when disconnected from their power source. Lights turning on and off is a common haunting phenomenon. Presumably then, if electromagnetic forces can be manipulated in a limited way by unpleasant entities in the metaphysical dimension to trouble us, then these same forces should also operate in the opposite direction to affect them. As yet we do not have any real evidence as to what might work, so I hope any individual or family with a regularly occurring haunting problem is brave enough to methodically run through some of these possibilities.

Astral Intrusions In The Physical Dimension

This is a vast subject covering many other aspects of paranormal occurrences along with ghosts and cryptids. Time slips and dimensional location slips should be mentioned here.

Some "lower" astral regions are inhabited by people just wanting to continue life as they knew it in the physical plane here on Earth. Old habits and lifestyles can be a comfortable place to stay with friends or family and minimal effort is required. They are free to do this and may carry on for hundreds or thousands of years as measured in physical Earth time. Although there may be some subtle pressure from your higher self to move on, free will is what it is all about and you can take your time. According to some of the OoB travelers they inhabit realms not unlike their homes previously on the Earth plane with towns, buildings and even cars, they were very familiar with. It would be what is known as a consensus environment with everyone agreeable and happy to be there. They will all be people at the same approximate level of soul development and reasonably comfortable with each other's company.

Time Slips

These are peculiar occurrences that have been described on numerous occasions and there is much literature on the subject going back even over centuries. Typically, an individual or a group travelling through an unfamiliar location, experiences something like a mist engulfing them or get the feeling of a heavy atmosphere, followed by a shift in the normal physical reality that surrounds them. Their new environment seems to acquire an aspect of an earlier time period, maybe going back decades or a bit further. The buildings could look Victorian and the people walking by are dressed in antiquated fashion styles. Our present-day travelers also realise that they can interact with people there and they may appear a little odd to the locals. They can even enter the time-slip buildings, exchange money and attempt to use their cameras or phones. There is a busy street in central Liverpool, Bold St, where this phenomenon has been repeatedly described. Those who experience this strangeness report that the street and its inhabitants seem to revert to a time-slip in the 1960s. Normally if there is more than one person at the location of the time-slip, then they will all experience the same event. After the event any electronic equipment used, like cameras will have no record of their use in the time-slip zone and even money that seemed to change hands disappears. Interestingly, none of the well-documented time-slips experienced by earth-bound individuals seem to occur in a future time zone, although this may be difficult to ascertain obviously, because the future has not been recorded and we have no data to compare with.

However, I would like to maintain here that I do not believe this is any type of time-travel. The physical universe is not constructed in a way that allows time-travel, as mentioned above, because the past and future are not places that exist; we live in a perpetual, rolling present. (However, visions of future possibilities but not fixed events, can be seen by some with that gift). This perceived shift in time is due to an anomaly where the local metaphysical dimension somehow intersects and bends towards the Earth's physical plane (see diagram below). This metaphysical warping originates from the lower astral realm and not from the Earth plane because the inhabitants of this realm have inadvertently shifted their consensus environment in a way that intrudes into ours. The people in this lower astral realm have recreated the lives and the places they were familiar with, while here on Earth, and sometimes we stumble into it with rather disconcerting consequences for Earth-bound travelers. Of course, any physical records such as photos and "solid" objects that were collected while in this anomalous zone, vanish when we return to our Earth-plane reality. The people existing in this particular lower astral anomaly can interact with us in the physical world, but we are both unwittingly, fully aware of what is happening at the time.

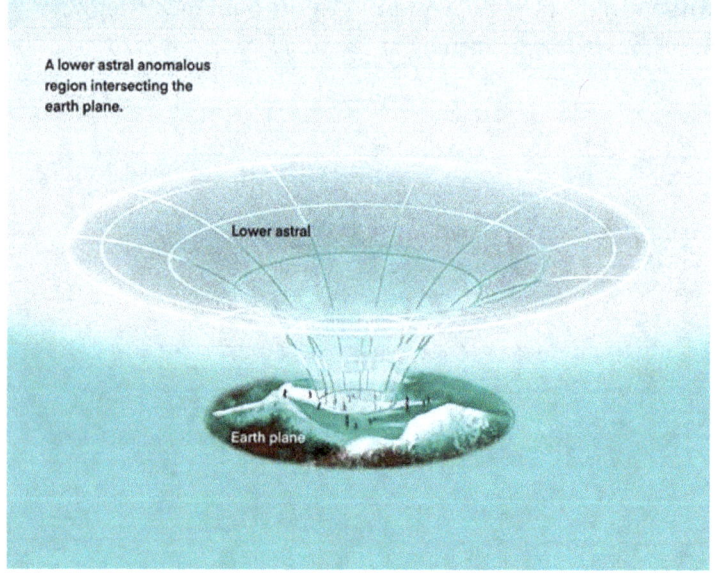

Figure 3. Time Slip Diagram

Predetermination

Accidents do happen. Pre-determination in your life cannot be absolute. If so, there is little point in our existence, we would be like musicians in an orchestra who have been given a score that is already written and we just obediently play the notes. I believe that our Creator would find this pointless: for Him and for us. If He already knows what we will do, then what is achieved by existing in a predictable universe as pre-programmed clones? Although this must sound blasphemous to all the religious types out there from many faiths, but I think even God does not know exactly what we will do, otherwise it renders our lives to an automated type of existence just following a pre-written script. Maybe He does not even know the coming lottery numbers. Although, some may be aware of Woody Allen's joke: If you want to make God laugh, then tell Him about your plans. Randomness is essential for the universe and God made it that way. But many people are unhappy that a benevolent Creator allows so much anguish in the world due to accidents or human wickedness. Our final destiny is to return to our Creator with spiritual growth and a story to tell; if He already knows it, then little or nothing is achieved. The analogy of an ocean-going liner full of passengers is helpful. God knows the departure point and the destination of the ship but what you do on board during the journey is not foreseen and is completely up to you. There is a two-way transaction here that enriches us all: our Creator explores and experiences a previously unknown and expanding universe. He also eventually acquires some company that does not consist of obedient robotic clones. Humanity, consisting of all the sentient individuals throughout the universe, who gradually evolve over many incarnations, will acquire a totally unique and enriched tapestry of existence. When we return ultimately to our Creator we will be like individual cells in the brain of God but keeping the option of an independence and of course cells that will never die. All the incarnating individuals in creation, like us, have taken the rocky road, a hard journey to spiritual development and fulfillment in a physical and unpredictable universe; but eventually it will put us ahead of the angels, if they really exist. We are in the school of hard knocks. Unpredictability is built into the fabric of our universe, in fact physicists say that it cannot work otherwise at the quantum level, the uncertainty principle is essential to the way our physical universe operates. We feel that our lives are special and unique and when we achieve something significant, we want this to last and not be something rather transient. All the details of our lives will be recorded and will be available for us to examine at any point we choose in the Akashic records held in the metaphysical dimension. However, the time and place of reincarnation *is* pre-chosen by you, or more precisely, you are nudged there by your higher self but after that, what happens here on Earth is totally up to you, with complete free will. We can imagine a scene where some higher ethereal beings were advising a young soul on their next Earthly incarnation. "You will be born into an Austrian family and they will call you Adolf. Your life plan will be to inspire a new political movement that will bring peace and prosperity to your country." What could go wrong? Obviously, it is absurd to think that our benevolent Creator wants to inflict great suffering onto millions of his people, so if things do go wrong, it was not written in stone that it must be that way, we have free will. Sometimes the circumstances of our birth situation can be fairly grim. It is often difficult to blame someone born in awful circumstances for the bad choices they make. All their decisions are based on choosing the lesser of evils. And what about the children born in war zones who have exceptionally brief lives or die very young of disease? I do not have a simple explanation as to why it has to be this way. Maybe we cannot be privy to all the arcane cosmic processes that guide our eventual spiritual destiny. But we must look at our existence in the totality of many physical incarnations and inevitably some of these will feel like a burden. It is worth noting that Jesus claimed that those spiritually identifying with him would be "saved." Somehow this spiritual conversion will reveal on a personal level a new type of awareness and an offer to individuals a short cut to reducing their karmic burden of future lives in the physical plane. This means we must reject our ego-centred lives and embrace the fact that we are part of an interconnected Cosmic family.

There is another aspect to future events that should be considered. It appears that OoB experiences enable some people to have glimpses of the future. In the metaphysical dimension there are different possible time-lines visible that we cannot see from the physical plane and there is a simple analogy we can use from here on Earth. At sea or ground level we have a visible horizon of about 5 miles, but if you ascend in altitude, then the higher you rise, the further you can see as the horizon moves out. Those who have passed over, with a certain level of higher advancement, are also able to see these timelines and sometimes advise those individuals who have near death experiences about their life options. Those people who are clairvoyant, also known as second-sight, believe they can see future events, for example the deaths of friends or relatives. Some lucky individuals have repeated dreams of the winning racehorse a day before the race. On a more theological level Jesus on many occasions told his followers what was going to happen, especially near the end of his life. He also said he would return in 2000 years and the book of Revelation in the Bible outlines many grim prophecies for mankind in the End Days that are prior to his arrival.

There is a fundamental problem with the future being absolutely fixed or predetermined. Many of the future visions that some individuals claim to perceive involve accidental events or personal tragedies. If for example someone with the second-sight "sees" themselves or others in say a plane or train crash in the near future, then in principle that person involved can go out of their way to avoid travelling by these means. Thereby preventing that accident from occurring, so that the future has been changed and is not predetermined. Free choice is the fundamental principle of our lives. So, when future visions are experienced, we must only be seeing possibilities or probabilities but not certainties. So, viewing possible events in the physical universe from the perspective of the metaphysical dimension does give the viewer a certain advantage in the timeline but it is not absolute. Over history mankind has had many prophets, most of them religious. If these prophets had been doing their job properly then they should always be wrong, which may be bad news for them and their reputations. Prophets nearly always predict trouble and pain so if we paid heed to them, those troubles should theoretically be avoidable, because we were able to change direction and dodge the impending calamity. As mentioned above a biblical apocalypse is predicted for mankind because of our wickedness. So presumably if we become God-fearing and righteous then this apocalypse should therefore not happen. Unfortunately, the direction that mankind has taken may already be locked in.

Occasionally good evidence that accidents can happen comes from many curious and slightly spooky occurrences. For example, the driver of a car speeding along a country road "hears" in their head a loud instruction to stop the car and pull over! Perplexed, the driver does so, only to experience a car coming the other way on the wrong side of the road just missing their stopped vehicle and thereby avoiding a catastrophic accident. Also, the story from a woman walking in the highlands of Scotland. She was hiking in a remote region when very quickly a dense mist descended so that her path ahead was almost invisible. She continued along what she believed was the right path but was suddenly confronted with what appeared to be a wooden bar across her route. So, she deviated from that path and continued her walk. On the return journey after the fog had lifted, she came to the point where she had seen the wooden bar but it was no longer there. However, she realized that what she thought was her original path would have led her over a vertical drop and certain death. In another interesting account, there were children at a church picnic. They were gathered at the bottom of a small hill in a field next to a lake but a local farmer had parked his tractor at the top of this hill and left it unattended. However, for some reason the tractor was not properly secured and it started to roll down the hill towards the children's picnic. One of the teachers saw what was happening and tried to warn the children of the tractor's approach, but they were preoccupied and seemed completely unaware of the danger. The teacher then spotted a driver in the cab of the tractor struggling with the steering wheel trying to divert the tractor away from the children. Fortunately, at the last moment the driver managed to change the tractor's direction and it came to rest, avoiding the children. The teacher ran to the tractor to thank the driver for his actions; but when she got there, there was no driver. Similar but more detailed accounts of assistance coming from the metaphysical realms in times of crisis were described by the author John Geiger in *The Third Man Factor* (26). He recounts many instances of individuals in desperate and extreme physical circumstances receiving help from mysterious, otherworldly people who may, or may not have been, wholly visible. Sometime the visitors would seem vaguely familiar like a departed old friend. Once the danger has passed these visitors vanish.

There is an important conclusion to be drawn from these types of personal metaphysical interventions into the lives of people on Earth. Why are they necessary? If the universe is a well-oiled pre- programmed clockwork machine, then any last-minute corrections should obviously be unnecessary, because everything that happens is already predetermined in the cosmos with no adjustment needed. Clearly not so. Furthermore, some people who have been in accidents and had near death experiences, have encountered advanced beings who told them that it was not their time yet and that they should return to their physical body and spend more time on Earth. Unpredictable things do happen and we have a choice.

Memory

The humanoid brain is probably the most complex, sophisticated and mysterious structure in creation. Maybe it is the seat of our soul. How it stores and processes information is still essentially unknown and rather controversial. Neuroscientists currently believe we have two types of memory, short term and long term and they function by different mechanisms. They claim our short- term memory is mediated at a neuro-electrical level and is capable of storing our experiences on a temporary basis. It is a crucial mechanism for storing relatively small amounts of information as we go about our daily life. The time of storage is up to about 1 minute. If our experience is significant because it is sufficiently emotional or repeated often enough, then this memory can be laid down in long-term storage. This consolidation of short-term memory into long term is thought to involve changes at a molecular or synaptic level across our brain neurons. Synapses are the junctions between neurons and if the signal across a synapse is repeated often enough then these connections become stronger or permanent. Eventually the memory associated with this synaptic change would become, as they say, "hard-wired". A simplistic comparison with a computer is helpful here. The short term or random-access memory (RAM) is stored on a micro-processing chip. Subsequently this information can then be sent to the computer hard-drive for long term or permanent storage. However, scientists are not sure exactly how biological memory works.

I would like to offer a very different hypothesis. The organic brain must certainly be processing sensory experience received initially but what happens next may well be very different. There is good reason to suspect that long-term memory is stored *outside* the physical brain on a metaphysical level. Not just in humans but for every living organism. A similar view has been considered by the scientist and author Rupert Sheldrake who refers to this phenomenon as *morphic resonance* and he has published widely on this subject (27). He describes morphic resonance as a non- physical field connecting all individuals within a species which provides them with a collective memory. He would go further and even include a form of memory existing in non-living material.

There is a problem in that the physical brain is finite in size but is expected to and probably does, somehow record every second of experience during our existence. For example, if our life span were many hundreds of years, then obviously a physical storage organ for memory would become "full", rather like a computer hard drive. How can a physically finite container store an infinite amount of data? Those scientists claiming that one day we might live to a thousand never address this obvious contradiction. So, there must certainly be a limit on the human life span if the brain has finite storage. We could imagine a form of Alzheimer's disease in reverse where early memories are wiped out to make room for new ones. But the prospect of forgetting our cherished childhood memories so as to make space available for new ones is certainly not appealing. However, the capacity for memory does seem to be infinite, although this has not been fully tested. Our recall certainly may fail as we get older but that is likely to be due to problems in the neurological mechanisms accessing or processing our memories, rather than storage. If information storage is outside the brain, then obviously there are no limits.

The probable location of our stored memory and unique experience is the Akashic Records. All information from human experience generated in the physical plane is thought to be stored in the metaphysical plane according to many well-known, informed sources like Edgar Cayce and Rudolf Steiner. This storage is like a Universal Memory or computer system with all our thoughts and memories stored in a cosmic equivalent of The Cloud, sometimes called the Book of Life. Only you have access to your own records unless you give special permission to someone else. As we can see from advertisements in the back pages of newspapers and on the internet there are many practitioners of Akashic Record divination offering a service to interpret your record for a fee.

Prodigies, creative geniuses and people with amazing numerical abilities like calculating the number Pi to 100s of decimal places in their heads, will not be doing this calculation inside their physical brains. This information is already available in the astral Akashic records so those who have special access simply read it off, rather than calculate it. In principle we all have access to this information but we are not sufficiently entuned or engaged with this resource.

The Akashic records may also be a record of past physical events in a specific location as well as an individual's personal actions and choices in the physical world. Some authorities on the Akashic records believe that recordings are made of all physical events in all locations which can be accessed by anyone with practice. It could be described as an astral hologram that can be accessed certainly by any person who took part in the event and probably anyone with "viewing" ability could see what past event occurred at any geographical grid reference. The implications of this, if true, would be astounding. Imagine being able to solve what occurred at a crime scene by tuning into the Akashic Record of that specific location. At present this would only be possible by individuals with that specific psychic ability. Some of the so-called hauntings of buildings and other locations seem to involve what might be described as a recording, on a loop, of the events that took place there. Especially where heightened emotion was present.

Strong evidence for memory storage outside the brain comes from people who have OoB experiences. Obviously if you observe events and are able to communicate with other people during OoB travel, then your physical brain is not there and cannot be participating. But those people returning after an OoB experience can remember, often very clearly, what happened without a physical brain being present, so how is that possible? Further evidence comes from memories of previous incarnations. Some of the memories from past life regression analysis of previous incarnations come from investigation under deep hypnosis so may be considered as controversial. However, the examples particularly from India, of children who died very young in a previous life and then remembering many details of it in their current life, are quite striking. So how are they able to recall events their current brain never experienced? Where is this information?

The amount of information stored in the Akashic Records from all civilisations in the universe is massive or possibly infinite and in principle we can all have access to it. But there is likely to be different levels of privilege depending on a person's state of spiritual attainment which restrict our access. A further possibility for non-physical memory storage is at sub-atomic or sub-quantum levels. As mentioned earlier, there is a vast amount of space between atoms and that must include inside the neurons of our brain. The brain may contain a non-physical component at the sub-atomic level that interacts with our physical brain in a way we do not currently understand. As mentioned before the relative distances between atoms in normal matter is enormous so there would be plenty of storage space. When the scientific technology allows us to explore space at a sub-atomic or sub-quantum level, we may discover and understand the "etheric" matrix that underlies creation and projects into our physical universe. This is the background where our metaphysical bodies have their permanent and original existence. Consciousness and memory may well reside here at its most fundamental level.

The Judgement

All faiths and religions affirm that after our physical death we arrive in a place where our recently completed life is assessed. The Bible is quite insistent regarding the Judgement and so is Islam. Buddhism and Hinduism are similar but believe that the process of judgement applies through the laws of karma and reincarnation. So, what really happens? As ever, not many passed on individuals come back to tell us. People who have had OoB experiences, especially a prolonged OoB experience can give us some clues. For some people an OoB experience is unplanned, like a near death experience (NDE) due to an accident or illness but for others it can be deliberately entered into. For those who had controlled OoB travel like Jurgen Ziewe or Robert Monroe a prolonged life assessment process was not experienced personally but they were shown, or allowed to visit, fairly grim locations, where rather hellish conditions prevailed after a "judgement" fell upon others. The author Lynda Cramer in her book *Five Years in Heaven* had a NDE after an acute serious illness. This provided her with the opportunity for a more detailed analysis of a life review and she described what was perceived as several years in the afterlife but was actually only a few days by earthly measurement.

The first step is that shortly after physical death a panorama of your life kicks in and the information revealed in this review has been described as watching a film where you are the main actor. Sometimes this may start while having a personal OoB experience and it can be interrupted. However, this is not the main event.

The next step seems to be delayed but cannot be avoided. The judgement process does not involve being confronted by a stern, grey-bearded old man pointing us upward toward a cloud where we can play a harp for eternity, or downward to a torment of fire while being prodded by demons with pitchforks. There is probably a lot of individual variation here but generally we are directed to an environment, probably a building, in our astral level where "higher beings" or guides invite us to view the entire experience of our just-completed life. All our experiences and life decisions are imprinted in the Akashic Records so we can access them at will in special places for re-examination. It is like watching a movie in which we take part as the main character and can be paused at any point. The important difference is we do not just "see" this movie but we also "feel" it from the perspective of other individuals we interacted with. The protective shield of our physical body is no longer available when we leave it, so the extreme personal anguish that we become exposed to during the review may be quite prolonged in some people with accumulated bad karma. We are able to perceive on a very personal level both the good and the bad consequences of our behaviour and choices that we make at any point in our life. It seems that going through this process has a significant cleansing effect on our personality as we forgive and are forgiven. In the higher astral levels any guilt, anger or other mental baggage is easily spotted by the people around us because we are much more mentally transparent. If we are unable to clear this type of mental blockage then our spiritual vibrational frequency is lowered and we remain at a lower metaphysical level. Possibly further physical incarnations with those people we have caused problems for may be required for us to move on if serious issues need to be resolved. Of course, many of our actions in physical life involving others were inadvertent or accidental, so it is our *intention* that really matters. The question many people want to ask is: What happens to the real bad guys? Like serial killers and genocidal dictators. During OoB travel Jurgen Ziewe was given guided access to a location where a recent suicide bomber had ended up and it was rather hellish. The perpetrator was trapped in a dark place in a pit underneath the moaning, writhing bodies of his victims from which he was unable to escape. Some religious clerics from his faith were there trying to help him but without success. One of them said to Ziewe when you get back to Earth you must warn the people not to do this. These "hells" are not for eternity but are difficult to escape. On the other hand, of course, there are also the saintly Mother Teresas of this world. No doubt they find themselves fast tracked to a more heavenly realm after death and may join a group sometimes called the Ascended Masters. There is also a system that allows you to bypass the worst experiences of the Judgement, as Jesus said, if you believe in me, you will be protected from this process. Some sort of celestial pardon seems to be available for those who have a genuine spiritual conversion. One day in the not-too-distant future we must hope that this will apply to everyone on our planet. Humanity will one day realise this life review happens to us all and we must modify our behaviour accordingly.

Synchronicity: Do Prayers Work?

A common thorny question is: do prayers work? If the answer was yes then of course everyone would be doing it. If you put the question into an internet search engine the response from individuals and "experts" is predictable: maybe yes or maybe no, so it depends. Theoretically prayers are not necessary as that part of you with a divine link and knows what is best for you, ie. your higher self, is already aware of your desires and intentions. This information is automatically passed onto an abundant and benevolent universe that provides for opportunities or meaningful coincidences to occur in your life, also known as *synchronicity*. This is not the same as predetermination as you are attempting to make a choice and free-will is an essential part of our existence. These events should bring people and places together for your benefit. The analytical psychiatrist Carl Jung (27) was one of the first to describe this phenomenon after investigating the Chinese text for divination, called the *I Ching*. Christianity certainly insists that collective prayer can work, especially if it is to the benefit of others rather that self. For example, if the vicar of a local parish is aware the church roof is about to fail, he may ask his congregation to pray for divine help in its repair. The prayer could be answered when he receives an unexpected legacy, made to his church, from a recently deceased parishioner. However, should this vicar pray that he can successfully seduce the beautiful wife of one of his congregation, then that prayer is likely to fall on stoney ground. Praying for things like the healthy recovery of sick family and friends seem to have rather inconsistent results as there may be karmic issues involved with an individual's wellbeing. There have been a number of statistical studies looking into the efficacy of prayer, again with inconclusive results but it appears saying prayers for others gives you a healthier, longer life, rather than receiving prayers or being prayed for.

As the universe expands and unfolds then people and events can somehow be tracked and diverted into a closer proximity for their benefit. This interweaving of our lives over time seems rather mysterious but with an extra dimension available, then a conscious universe can direct this process. Of course, in principle the opposite could occur to cause harm but the universe only works for the greater good of us all. The enormous changes that are coming soon to this planet will have a profoundly uplifting effect on humanity as we will live in harmony with a universe, that produces a flow of generous synchronicity in our lives. We will all be aligned with the universal consciousness that very few of us are currently aware of, where prayers may no longer be necessary and even religion will become redundant. It will be like having an extra sense to the five we already have. I outline in the next section how these changes may take place.

THE COMING PLANETARY CHANGES

Apocalyptic Possibilities

Many sources from New Age theories to religious texts tell us that profound and exciting changes are soon to affect humanity on this planet. There will be a flowering of consciousness and a new awareness. These changes are sometimes described from an astrological perspective or in religious texts, for example the Bible in chapters from the Book of Revelation predict and describe the so-called end-times as the Great Tribulation. Other religions talk about massive events prior to a new age including the Koran with a description of the Last Judgement. What this means in reality is that mankind is to be made aware of its purpose, its origins and possible futures in some way that leaves no room for doubt. Our spiritual heritage and what happens to us after physical death will become clear. Also, how important our lives are in the physical plane in relation to long term spiritual progress. Exactly how these changes will be manifested is not clear and in the Bible for example, the run-up to the changes is described as being fairly dire for humanity. A number of apocalyptic scenarios are described like Armageddon wars, fire from the skies and gigantic earthquake upheavals. Many Biblical scholars have tried to interpret the texts but there seems to be no definite consensus unless you want to believe the fire and brimstone preachers where most of the human population on Earth will be wiped out. The Bible seems to take the view that events which will take place are somehow contingent on mankind's behaviour. The logic of this, of course, is that if mankind suddenly became virtuous and God-fearing then such awful events would not be necessary and will *not* happen. Personally, I wouldn't count on it but I realise predicting catastrophes has always been a popular theme throughout human history, especially from theologians. Other interpretations believe there will be a more gentle change. However, looking at it from a realistic and slightly gloomy point of view in Nature, when big changes take place, this would normally involve a new start after wiping the slate clean. For example, the end of the dinosaurs came after a giant meteor strike on Earth and there followed a new path for evolution. There are 3 main apocalyptic scenarios and they have been well covered by Hollywood blockbusters with stunning CGI graphics and 3D, so we know only too well what it may be like. Firstly, an event originating externally to the planet like a meteor or comet strike. Films like "Deep Impact" and "Armageddon" came out in the 1990s with mankind threatened but then saved by Bruce Willis. Secondly, forces are unleashed from within the planet like the earth's core heating up and producing gigantic continent-changing earthquakes which wipe out most of humanity in the film "2012". This date was thought to be the end of time in the Mayan calendar and prophets were telling us this would be the end of the world. The third apocalyptic possibility is the product of man's doing, like a nuclear holocaust. There are many films about humanity struggling in a post-nuclear apocalypse wasteland like in the "Mad Max" series. Of these options only the nuclear apocalypse is totally dependent on mankind's reckless actions and would therefore justify warnings of doom from prophets if mankind does not change. If there were to be an apocalypse then that would seem to be the most likely possibility. In the distant future of our planet, I fear that the 21st century will become known as the time of the Great Destruction.

We must also realise that prophets and religious leaders do not like letting a good disaster go to waste. Noah's flood is a good example. Paleontologists and geologists have very detailed hard evidence that Noah's flood was caused by the rapid filling of the Black Sea about 10,000 years ago. Around this time large masses of ice were melting as the Earth emerged from the last ice age and flooding was common around the globe with rapidly rising sea levels. The region now underwater in the Black Sea was a large land area at a low level and protected from the Mediterranean Sea by a land bridge across what is now the Bosporus Strait. Eventually the rising seas broke through this land bridge with sudden widespread catastrophic flooding over a vast region and no doubt there may have been heavy rains at the time to consolidate the Bible story. A wise man of the time called Noah may have realised what was going to happen and took precautions by constructing a large boat. This was an event that was going to happen regardless of whether mankind was wicked or not. But if you are a prophet or religious leader it may well improve your authority and credibility with your flock if you convince them such disastrous events happened because they ignored your words of wisdom. The consequence of course was that God punished them for their wickedness with a deluge. The Bible describes punishing the whole world with a flood but of course this was a regional event. Similarly, if there were another apocalyptic event described as affecting the entire world in the end-times we should be cautious. This narrative of end-times in the Bible may only refer to apocalyptic events occurring in the region of the Middle East where all authors of the Bible originated from.

If nuclear weapons are ever used again the possible scenarios for their deployment will certainly have been thoroughly explored by military strategists and security services in most developed nations. The odds are that the Middle East is one of the most likely places to see nuclear weapons used in anger because of the long- term Arab resentment towards Israel. Relatively tiny Israel is surrounded on all sides by hostile states with large populations who would like to see its eradication and not surprisingly has resorted to maintaining nuclear weapons for self-defense. It is also known as a "one bomb" country, meaning that it is so small that a single nuclear detonation is likely to incapacitate the entire nation.

Delivering a nuclear strike by using a bomb on the end of a missile must be one of the most inefficient and wasteful approaches for such an attack. Just consider the massive resources a small country like North Korea is putting into its nuclear missile programme in order to justify international credibility. Smuggling a nuclear device into your target country and leaving it in the back of a van parked downtown is infinitely easier, and Israel, or more precisely Tel Aviv, is likely to be hit this way. A possible sequence of events could unfold as follows. An anti-Israel terrorist group manage to get hold of a smallish nuclear bomb and explode it in downtown Tel Aviv. The destruction and casualties would be horrific. The Israeli nuclear arsenal is almost certainly well dispersed around the country or elsewhere and would be available for any retaliatory nuclear strikes. The Israeli military high command must have thought through this scenario and their response is likely to be merciless and they may well take the attitude that if you have nothing left then you have nothing to lose. Any of the nearby countries supporting anti-Israeli terror groups are likely to be hit. Tehran in Iran would certainly be struck and possibly Riyadh in Saudi Arabia. Other countries like Syria and Iraq have already self-imploded and are likely to be spared. There are many other middle east countries although not actively hostile, do regard Israel rather like a festering sore and would be acquiescent in its destruction, like Egypt and Jordan, and they may be considered targets. The Israeli military would be very keen to know where the bomb came from and if they find the fingerprints of Pakistan or Iran all over it, as is very probable, then things could get much worse. Islamabad and other cities with military links in Pakistan would be probable targets. Pakistan as a nuclear armed power is likely to retaliate and then all hell breaks loose. It is difficult to know if other countries would be dragged into such a conflict because of their support or hostility towards the various combatants. But nuclear weapon exchanges between Israel and much of the Muslim world is going to be bad enough. Millions will die and many regions will be left uninhabitable. Israel has in all likelihood already informed the rulers of places such as Iran that if a bomb is detonated in Israel, then they will hold Tehran responsible because it is well known for supporting anti-Israeli movements. Tehran would have been told that retaliatory strikes against their country will be automatic after any nuclear attack on Israel. If such a devil's pact has been made, we cannot be sure but a particularly ruthless terror cell with access to a nuclear weapon may not care anyway.

The scenario outlined above is not unrealistic and would result in a regional Armageddon, leaving the rest of the world spared from the immediate damage. It would fulfil much of the Biblical prophecy about huge destruction taking place in the end-times involving the holy lands where Israel and many of God's chosen race are destroyed by fire from the sky.

There is another rarely considered side effect of nuclear detonations on this planet or in near Earth space. Apparently, according to ET communications (the Sirius Disclosure project) and the clairvoyant Morris Bradley that such enormous explosions can rip through the fabric of local physical space/time and cause a great disturbance to the adjacent lower astral dimension. This would seem reasonably plausible considering the massive amount of energy released into the immediate environment when matter is converted to a huge amount of energy. This massive energy release causes some type of disruption that ripples through the nearby metaphysical levels which can have unfortunate consequences for the occupants residing there.

The Dead Man's Fuse

This is another scary possibility, probably already implemented. During the time of the cold war both sides would have realized that delivering nuclear weapons by missile or plane is very inefficient. One retired US army colonel, Tom Bearden, has stated publicly that the Soviets almost certainly smuggled small nuclear devices into target cities of interest, eg. New York, Los Angeles and London. They will be well hidden, maybe in domestic dwellings with underground basements or guarded lockups. Originally these weapons would have required detonation by a local operative in place, who would obviously perish in the blast, hence a dead man's fuse, but now this could probably be achieved remotely by electronics. It would be very naïve to think that Western military commanders have not done the same thing, to what were then Soviet cities like Moscow or Murmansk at some time during the Cold War. Smuggling in such weapons is unlikely to present a huge challenge, as all these nations have enormously long borders and coastlines that in places are very leaky and poorly guarded. As the military powers on both sides are aware of the situation, then the principle of MAD or mutual assured destruction still applies as with ICBMs or plane-delivered nuclear weapons. It is frightening to think that the citizens of these major cities could be living just a few metres from a live nuclear weapon and be completely unaware of it and there is no early warning system with such a weapon placed in-situ. Trying to look at this positively, then the stalemate created means neither side is likely to pull the nuclear trigger and the tactic of attempting to launch all your missiles first is rendered useless. However, it is a precarious form of mutual blackmail and the cock-up theory of human endeavor means if it can go wrong then it probably will. Although the Soviet Union no longer exists, I doubt if either side is trusting enough of the other and such weapons would still be in place, especially with the current

Russian government aggressively attempting to rebuild an old empire and its warmongering in Ukraine. Who knows, in all probability other nuclear armed nations may have done something similar and we might never find out who planted the device. Obviously, should senior authorities in the military ever be questioned about this in public, then of course we would receive an outright flat denial, so there is little point in asking and we should assume the worst.

Avoiding Armageddon

It is possible humanity turns away from the path of self-destructive warfare and the above possibilities are avoided, so what happens then? There are many scholarly interpretations of Biblical text as to what is meant by the second coming ie. whether it is a genuine reappearance of Jesus or a metaphorical return. If Jesus is to return physically as an adult in a new reincarnation, then he is probably alive now but not necessarily on this planet. And for the second time around he is supposed to arrive in a blaze of glory rather than from a humble manger. The second coming may be a metaphor for the transition to a greater awareness for humanity of the non- physical aspects of our existence. This change must also be almost instantaneous for the whole planet and not a gradual, piece-meal process. The population of an entire planet must experience the change together. So how will this be manifested? A likely possibility is with the assistance of Extra-Terrestrials described in a later chapter. However, be prepared. If an Armageddon should come, then make sure you are not living in a large city.

The new awareness is sometimes described as a lifting of the veil that hides the metaphysical dimension from us. What we are likely to experience are manifestations from higher dimensions becoming visible to everyone; maybe in the sky or all around us and a more accessible spiritual awareness revealed to those who seek it. So-called miracles will become normal events. Physically deceased friends and relatives should be visible and able to communicate in some circumstances. And the process would be two-way so that those who attempt out of body experiences will find it much easier. They will be able to visit astral planes from where they came and those places where they would like to continue development. Hopefully the more powerful astral light will have a cleansing effect on the lower regions where the more unpleasant entities; human and otherwise are residing. The black clouds of psychic negativity emanating from these lower astral regions diffuse into the physical and have had a detrimental effect on humanity for many an age; subtly affecting our decisions. It will be like draining the swamp. The OoB travelers report there is no higher supernatural being created by the Source who has fallen, now in spiritual opposition, that we have named Satan. But whether Satan exists or not, preachers and religious leaders by necessity, have had to create him. When a person does something criminal or evil, we prefer not to blame the individual but rather claim this person is not really responsible for their actions; they have been led astray by evil supernatural forces beyond their control. In this way we can avoid responsibility for our own behaviour; we find someone else to blame and Satan is the perfect scapegoat; we always need to find a bad guy. Christian preachers tell us that Satan's plan is to deceive mankind into believing he does not exist. This is nonsense, perpetuating belief in a devil is compulsory for all religions even when such a being does not exist, in order to justify, highlight and personify the good/bad polarity in our everyday choices. However, there is a real Satan but he is not an external entity. This Satan is the uncontrolled human ego that demands power and unlimited personal gratification. It chooses to ignore or resist the harmony of a balanced and conscious universe where all life is interconnected. In extreme cases it can be responsible for bringing devastating damage to humanity. This potential exists in all of us and we have free will.

Experiencing a Divine awareness or cosmic consciousness is not as simple as flicking a switch. It seems quite impossible for most of us and many people spend a lifetime in worship or meditation in search of this goal without success. I suspect the problem must be the sincerity of our decision and achieving this status is quite a privilege and the key is looking within ourselves. A probable short cut to experiencing this awareness is being "born again," as described by many Christians who turn to Jesus. Ideally the role of religion or the church should be to help their congregations find this awareness because communal worship has a powerful effect on the individuals taking part. The church should be functioning as a lens, focusing the divine signal that is broadcast throughout Creation then beaming it towards their flock. Unfortunately, on this planet certainly, most religions have lost their way. They have become obsessed with power, ecclesiastical frippery and often vast wealth. Religious leaders are typically preoccupied with the politics of their religion rather than saving your soul. Consequently, religion ends up not being a lens for Divine awareness but an umbrella blocking the Light.

How other changes are implemented we can only speculate on. For example, the Bible describes the new age as being a heaven on Earth where the lion will lie down with the lamb but this may be only a metaphorical description. There have been billions of years of evolution on our planet resulting in nature which is red in tooth and claw and a food chain with top carnivores predating on other animals. It is hard to see how divine astral energy will alter the basic physiology and biochemistry of every living creature on our planet and cancel out eons of evolution which has produced the delicate balance of nature between millions of species including mankind. But this type of paradise may work in the afterlife.

Extraterrestrials (ETs)

When the extraterrestrials show up for real their first message will be: there is a God and you must believe. We can speculate on how this message will be delivered; some ETs are extraordinarily telepathic and we could be informed in such a way. Possibly they could hijack our media or internet to broadcast this message. Their presence here and their paranormal abilities will likely be evidence enough. In fact, some of their abilities are so profound that they make Jesus's paranormal achievements look pretty ordinary. Currently most Earth-visiting and advanced ETs regard us with some pity and often amusement as we, unlike them, are almost totally unaware that there is an eternal non-physical existence around us where we will eventually reside. Our behaviour would change so much if only we realised. Planet Earth is regarded as a rather backward inhabited planet, definitely in the remedial class or even barbaric, by intergalactic standards.

Advanced ETs will be very aware of the laws of karma and spiritual progress and they are not coming here to attack or take over humanity; they are inherently altruistic. Some high-profile commentators in places like the military leadership have expressed concerns that ETs are potentially invaders. This is lamentable and shows quite confused and paranoid thinking. If, as many suspect, ETs have been observing and interacting with our planet for thousands of years, then why did they not invade in the past when this would have been so easy? Why wait until now when this planet's population is in the billions rather than millions and we have industrialised warfare and nuclear weapons. Many ETs will have cleared karmic problems from their existence over many lifetimes and such actions would have personally disagreeable long-term consequences and they look at our planet as one that requires urgent assistance. All inhabited planets have a special place in Creation for the spiritual advancement of their inhabitants and cross-reincarnation can take place for individuals between different planets who require different life experiences. Planet Earth has come to a critical point in its development and ETs are here to help. Occasionally people abducted by ETs (normally the Greys, discussed later) who were not psychically sedated, complained to their abductors they had no right to carry out these investigations and examinations. Curiously, ETs replied that they *do* have the right to do these things. This suggests that there may be an over-arching level of responsibility that these advanced beings must have over more backward planets and we are currently quite unaware of this. ETs do not want to be placed in the position of gods when they come here or on other more backward planets. They are human of course and the only difference is that they have been around much longer. We must also understand that ETs are not here because we have something they want, like valuable minerals or genetic material: a popular theme with ET invasion scenarios in the cinema. They are so advanced they can make anything they wish, probably within minutes using specialised trans-dimensional technology and will have access to the almost infinite supply of zero-point energy. We have been seduced by Hollywood science fiction that ETs coming to Earth will inevitably be hostile. In fact, even some of our most brilliant scientists have been very concerned that we are attempting to contact any ETs, eg with the SETI project (the Search for Extra-Terrestrial Intelligent life). It is surprising that so many highly intelligent people in science have been overcome with this paranoia of invasion from extraterrestrial aliens. Even Stephen Hawking believed we should not advertise our presence to the wider universe. He was troubled that if ETs discovered our planet, it would be disastrous for humanity. He pointed out that if we look at the history of colonization on our own planet then the indigenous populations inevitably seem to get a bad deal at the hands of an incoming colonising people. Especially if we draw a comparison with the first-world powers spreading around the globe during the 19th century, it was normally bad news for the locals. For example, look what happened to the South American Incas and Aztecs after the Spanish arrived in the 16th century. Compared with humanity here on Earth, ETs are highly advanced spiritually and intellectually and will realise that when they arrive here, they have a delicate balance to maintain, as they do not want to look like gods or invaders. They will bring us the message of peace and brotherhood that exists throughout the universe and it has not quite reached here.

The Greys

The ETs that feature predominantly in human interactions are known as the little Greys because of their appearance and are particularly psychic. They can read our minds easily and have an ability to project telepathic movies into our heads. Other effects include triggering out of body experiences, the life review process and an immediate loss of consciousness. Clearly, they have an ability to manipulate our astral bodies using properties of the metaphysical dimension that we currently do not understand. There are also the physical effects: many are observed levitating (hovering off the ground, like Jesus walking on water) and moving physical objects mentally. A number of close encounters have been associated with the rapid healing of injuries or ailments. Most of the close encounters experienced by observers claim that mind altering telepathic effects were common during their encounter, especially in the so-called abduction cases which predominantly involve the little Greys.

The Greys as a race seem to have a kind of hive mentality. They are collectively able to share the same experience of a single individual if they wish, by a telepathic link. For example, one individual viewing an object can share that vision with another of their race almost immediately, even if they are separated by a distance of light years. These visions can also be shared with one of us on Earth if they choose. This telepathic connectivity is one that we may find rather disturbing but certainly in the higher afterlife realms such interaction is likely to be a normal ability. What the Greys have achieved is to bring this level of communication into the physical plane. We can speculate that this sort of psychic development could be described as progress, so maybe many generations into the future we will also wish to become less individualistic and be more communal as a race. Subsuming our personal identity for the sake of a soul group is a step most of us are not yet ready to take on this planet, as our individuality is considered very precious. But maybe after much spiritual advancement this becomes more appealing as we recognize that forming close bonds with kindred spirits is the way forward.

It must be pointed out that not all ETs are warm and friendly. In the early years of UFO and ET encounters, US military officials who were in-the-know allegedly put ETs into four broad categories: psychic and friendly; psychic not friendly; non-psychic but friendly and non-psychic not friendly. However, this did not mean that the non-friendlies were actively hostile. Many people who have reported close encounters or abductions felt they were not treated particularly well and there may have been a few deaths which were probably accidental. After being returned, some abductees have reported radiation-like burns, obscure health problems and nightmares. The little Greys reportedly seem rather cold and dispassionate in their dealings with humans. They are simply doing their job and we are just a rather tedious and backward race. However, they are normally overseen by much taller, similar looking beings who radiate a telepathic benevolence toward their human abductees. The purpose behind the abductions is not clear but often involves some type of medical investigation. Presumably these must be done prior to the great changes that are coming to this planet. They could hardly perform these, sometimes rather intimate investigations on us, when we are fully aware and they would have to ask us first. Another reason might be that the selected abductees are being "psychically" primed in preparation for possible apocalyptic conditions that will affect the whole of our planet. These individuals will be telepathically woken at such a time and be required to act as guides or marshals to lead large numbers of people to safe zones or ET rescuers. Abductees have reported they were telepathically informed that at some point in the future they would be required to act as guides for a large crowd.

ETs also seem to have a sense of humour and winding up the Earthlings is an occasional sport for some of them. Some of the airplane encounters with UFOs suggest that ETs sometimes like to have a bit of fun with our pilots, who end up of course, being scared witless. Small UFOs flying around aircraft in circles and even nudging them sometimes have been reported. And there have even been reports of passenger airliners being deliberately dragged slightly off course by some kind of tractor-beam technology. The ET's long-term plans for interacting with us remain obscure and ambiguous but they manage to hide in plain sight thanks largely to government coverup and misinformation.

Unfortunately, quite a number of well-informed writers on this subject have taken a very negative stance against ETs. For example, Nick Redfern who has contributed much in the literature and on TV shows. Dr David Jacobs who is president of the International Centre for Abduction Research; Whitley Streiber the author of *Communion* and *Transformation*, who did not believe his encounters and abductions were positive (29). Also, Timothy Good who has written widely on the subject of UFOs for many years. I do not fully understand their concerns. If ETs were coming to conquer and subjugate the Earth, they would have done so easily, centuries ago. However, the human race has an instinctive fear of the unknown and of course some ET beings have a rather weird, even grotesque appearance, coming from another solar system, which makes us automatically recoil. The history of humanity with our constant fear of invasion, often with good reason, has meant that we assume all strangers are a threat.

However, a few abductees do regard their encounters as positive and even claim they are ET Grey/human hybrids including Judy Carroll in *The Zeta Message* (30). As someone who comes from a professional background in genetics, I am very skeptical about these claims referring to a physical genetic hybrid, because such a cross would not produce a human who looked anywhere near normal. Such a hybrid being walking around the streets on this planet would immediately attract much attention. The Grey phenotype (physical appearance) is so different from ours and it is very unlikely that a hybrid could be produced outside a laboratory, but of course ETs may be able to do this using an artificial womb. The physical structure of the Grey anatomy which has a very large head with a very thin body and tiny hips would certainly be unable to participate in a normal childbirth. It is also likely that ETs do not even have the same number of chromosomes as we do (that is 46). There are typically problems with a chromosome number mismatch when different species are crossed because the developing fetus will not survive as cell division is compromised. However, there are some cases which do work, for example horse and donkey cross that produces a mule, but the offspring would be sterile. I suspect the term alien hybrid could really mean that the people who make these claims have reincarnated on Earth after a previous life on different planet, presumably the Greys' planet. I am sure such cross reincarnation between planets does occur, but the gulf in metaphysical or spiritual development between two such races must be huge. The UFO investigator and author Dr David Jacobs claims that genetic hybrids between the Greys and humans are responsible for some humans having prodigious telepathic powers like the Greys possess. I am very skeptical about this, if only telepathy were as simple as a few genes. Far more probable is that any human who remains in the company of other telepathic beings for any length of time will also gradually acquire that ability. Telepathic linking is a property of the metaphysical mind which must be used as this is normal communication in the astral dimensions. Telepathy transcends any physical constructs or barriers. A frequent consequence in those people coming into contact with ETs are the after-effects of psychic phenomena unfolding in their lives subsequent to their encounter. Most Earth- based humans are relatively free from psychic abilities but exposure to extremely psychic individuals like advanced ETs, for even a brief period of time, somehow realigns or awakens our psychic potential. These forgotten or undeveloped latent abilities within all of us, become switched on. Such people may discover a raft of paranormal abilities, including some that are less desirable, like attracting lower astral entities and poltergeist activity into their homes. There is some fascinating reading on these phenomena to be found in the literature published surrounding the investigations at the Skinwalker Ranch in Utah. It is also interesting that ETs do seem to have manufactured some physical devices that respond to human thoughts, including their spacecraft. Possibly these machines include some biological neuronal tissue in their manufacture which responds to thought.

ETs like the Greys have a much reduced and basic body form and they are all so remarkably identical that they are obviously cloned as the process of normal childbirth would not be possible with their body anatomy. Such a race must have lost any interest in the lusts of a physical body eg. desires of the flesh like physical appearance, indulgence in food, drink, sex and physical exhilaration. They must be many reincarnations ahead of us so the trappings that are often associated with a physical existence are no longer of much interest to them; they have moved on and freed themselves of any karma in the process. Their bodies are basically a brain on legs and a large brain too by all accounts.

There is an interesting aspect to event prediction that should be mentioned regarding ETs. These advanced humanoids must be well aware of the capability for future visualisation of probable events from the viewpoint of the metaphysical dimension. If so, then why have there been supposedly a number of UFO crashes, especially like that at Roswell? Surely the highly evolved ETs would psychically sense an impending disaster and take steps to avoid it, especially seeing that a number of ET deaths allegedly occurred from these crashes. Their ability to see the future probabilities may not be much better than ours.

A number of other ET types have been frequently observed in close encounter interactions. For example, the so-called Nordic types who have long blond hair with beautiful features who are reportedly very benevolent towards us. Others are the Reptilian types, not so friendly, and a wide range of other shapes and sizes, but essentially all of humanoid appearance. There have been occasions when people have encountered an unusual individual, potentially extraterrestrial, who seemed almost indistinguishable from us, eg: Caucasian looking men with reddish hair and beards. If their spacecraft had not been visible the only give-away that they were not of this planet, was a strange paranormal interaction not possible with Earthbound humans eg, a particularly intrusive and overwhelming telepathic interaction or controlled psycho-kinetic activity. I have a suspicion that these individuals were actually Earth-born but had spent many years on an ET planet and in the process acquired many of their abilities. Rather like the scene at the end of Spielberg's film *Close Encounters* when the abducted airmen were returned. It seems throughout the universe that intelligent life has evolved based on the same model, ie: a body with a head, two arms and two legs, although the number of digits may vary. A comprehensive list of all the alien types can be found in *Real Aliens, Space Beings and Creatures from Other Worlds* by Brad and Sherry Steiger (31). Maybe when God created Man in His own image this applied throughout the universe and creation. This similarity or convergent evolution could also be explained by the theory of panspermia, originally put forward by Professors Hoyle and Wickramasinghe. If this were the case, then life did not need to evolve separately on each planet but was only required to evolve once in the universe. Small amounts of genetic material ie, probably microbial nucleic acids, were blown off by high atmospheric winds, or knocked off by impacts, from one planet and then are carried throughout interstellar distances. This should not be a great difficulty if billions of years are available for this process to happen. If interstellar space is teeming with primordial organisms, then these would rain down everywhere, including planets with suitable conditions for more complex life to evolve. The universe is around 14 billion years old so this would not be a problem. Some of our scientists are already testing the resilience of small invertebrates exposed to the harsh conditions in space and their survivability is impressive, especially the "water bear" or tardigrade which is a particularly hardy little multicellular creature only visible with a microscope. They can close down by switching to a condition of suspended animation which excludes most water from their bodies enabling long term survival in conditions quite impossible for normal life like a vacuum, intense radiation and cold. An Israeli mission to the moon crashed in 2019 which contained a sample of these creatures to establish their endurance. It will be interesting to visit this crash site on a future manned lunar mission to see if any have survived.

The writers of Star Trek may have got a few things right. One important rule for the starship crews was the Prime Directive. This required total avoidance of any involvement in the affairs of pre- warp-drive planets. ETs will have been watching us on planet Earth for thousands of years and may have undertaken a similar policy. The development of intelligent life on all planets must find its own path and freedom of choice is a golden rule throughout the cosmos. When technology on an emerging planet delivers the science behind anti-gravity propulsion and the equivalent of warp drive (another Star Trek prediction) then obviously the situation must change and things may have come to a head in recent decades on our planet. Interestingly, the creator of Star Trek, Gene Rodenberry, was present at the table in the 1950s when discussions were taking place over what to do about the UFO phenomenon.

It is a bit of science fiction fun-fantasy to imagine these things, but there must be a local galactic council meeting of advanced races who meet to discuss any problems. So, planet Earth will have been setting off alarm bells in recent decades with our developments in sophisticated mechanised warfare and nuclear weapons. A few belligerent and reckless individuals now have the capability to wipe out humanity. Such despotic tyrants and warmongers sadly are running many nations, some with nuclear weapons. These individuals cannot be permitted to block the spiritual progress of maybe billions of innocent people on this planet by triggering thermonuclear warfare. Although this may be an exercise of free will, there will be a point reached when such a consequence is not allowed as it affects so many blameless people. Currently planet Earth is likely to be under some sort of quarantine although most of us would be completely unaware of it. For example, if one of our nuclear armed countries wanted to explode an atomic bomb in space or on the moon, it would be prevented from doing so. I believe that if this were attempted then whichever ET group was monitoring our space activities would make sure the device failed or was deviated from its target. There are a number of unofficial reports from the astronauts who landed on the moon that they could see distant strange craft that seemed to be observing them. Of course, NASA would dismiss this information immediately and deny the reports with mundane explanations and misinformation. There have been many investigations into this type of report and have been highlighted in TV investigative documentaries like *The Unexplained Files* and the series *Ancient Aliens*.

There are some organisations like SETI who are constantly scanning the cosmos for evidence of communication signals from extra-solar ET civilisations and they are rather bewildered when they find nothing of real interest. This problem is known as the Fermi Paradox where in theory there should be many extraterrestrial intelligences just in our galaxy but none are detected using radio-telescopes. But even our brightest scientists just don't seem to get it; advanced ETs do *not* use technology involving electromagnetic radiation, including radio waves to communicate, for the simple reason it would be hopelessly slow.

Our scientists must abandon the Einstein dictum that nothing can travel faster than light. If you consider the age of the universe then there are probably many ET civilisations millions of years ahead of us and the light speed obstacle will have been overcome. If ETs wish to communicate with their home worlds hundreds of light years away, then they are using something else and we have yet to work it out. It will be along the lines of the "spooky action at a distance" as referred to by Einstein; it maybe a type of quantum entanglement or the Universal Consciousness. There is a curious possibility that arises from the ability of ETs to manipulate the metaphysical vibration of their space craft and occupants in order to take a short cut when travelling interstellar distances. This method of travel effectively dodges the problem of light speed being a limiting factor in the physical dimension. When ET employs this trans-dimensional shift device it means they are moving through the lower astral regions in what is essentially a physical machine. Obviously this means they are moving through some of the places we might occupy after our physical body dies. If this is the case, then meeting deceased friends and relatives in a machine that can take us to such regions would become a realistic possibility and the world would certainly change forever.

Antigravity

Currently our top physicists are very unhappy with the concept that objects with mass, like spacecraft, could travel faster than light and this is a big stumbling block when it comes to their belief that ETs are already visiting us. Einstein's famous equation is brought out to explain that any objects with mass would require an infinite amount of energy for it to reach light speed. Obviously, there must be a way around this and ETs will have worked this out, as they are coming to us from many other solar systems and even galaxies. Some scientists here have suggested rather ludicrous technologies to travel large astronomical distances like sunlight driving space- sails and ion-drive motors. These have no chance. Clearly ETs are using a type of space-time warping device. This means that matter itself is not being moved through space at a speed faster than light, which would contradict Einstein's equation. However, a bubble of space-time enclosing an object would be able to do so, because a small parcel of "space" moving through space-time is not restricted by these laws. We can only speculate as to what the limits on such travel speeds might be, but in order to travel between stars in a realistic period of time, suggests speeds of at least 100 times faster than light would be necessary. The full understanding and manipulation of gravity has been particularly elusive to our scientists, almost to the point of considering deliberate obstruction.

The American physicist Thomas Townsend-Brown during the 1920s observed there was a connection between electromagnetism and gravity when doing some small-scale laboratory experiments. Mainstream science largely ignored these observations because there was no theoretical basis behind them, certainly according to Einstein anyway. Townsend-Brown theorized that a positive charge created a miniscule gravity "well" and negative charge an equivalent miniscule gravity "hill". When these charges were separated by a capacitor then a slight movement of the charged object or gravitator, could be achieved "down the hill and into the well" in the direction of the positive charge. Provided there is a massive charge difference that is separated and maintained then the gravitator or a spacecraft will continue to move in the positively charged direction. He also observed the amount of gravitator movement varied slightly according to the time of day and seasons. He concluded that there was some cosmic scale interference, possibly high frequency gravitational waves emerging from our galactic centre that affected his measurements. He also noted that certain types of stone seemed to affect his results. If there is a link between gravitational waves and some types of rock, then this may explain the mysterious purpose behind some large stone structures like the pyramids and other unusual megaliths. These structures may be able to mediate a transduction process between gravity and electromagnetic energy. A number of scientific laboratories of his time attempted to duplicate Townsend's experiments but concluded: nothing to see here. The problem of course is; if there were something to see, ie that electromagnetism did affect gravity, then this would be very profound scientifically with huge military implications. So, people high up within US defense departments would be extremely interested in keeping this knowledge to themselves. Therefore, a good way to divert any further scientific investigation, especially taking place in hostile nations, would be to dismiss Townsend's work as unsubstantiated and not repeatable.

This subject has been comprehensively explored and published by the physicist Paul LaViolette (32). He has theorized that a new type of physics must be developed called sub-quantum kinetics which would provide a theoretical background to the principles behind antigravity and include modified classical physics. He has attempted to define the physics that operates at a much finer level of resolution in our universe, than is currently acknowledged. The antigravity devices produced by ETs are able to extract the almost limitless and free zero-point energy, supposedly available according to our theoretical physicists and use it to power the incredibly powerful electrical charge separators on such craft. Our scientists have been very slow to twig (or silenced in some cases) that there is a link between electrical charge and gravity. If anyone out there is competent in electronics and has a workshop, they might try building an antigravity device. Construct an asymmetric highly chargeable capacitor and fix the positive end to a large metallic surface and see what happens. It could be spectacular if you are lucky and in that case, you might also get a visit from the Men In Black as this technology *mus*t be kept secret. A number of garden-shed inventors may well have stumbled on these things but have been effectively silenced when trying to exploit them. Either by large amounts of money or more sinister means. For example, the Latvian stone mason, Ed Leedskalnin who built the Coral Castle in Florida single handedly, was able to move 30 tonne stones without massive machinery. Mysteriously, all his construction and garden equipment vanished quickly after his death. The secrets of antigravity must have been available to some ancient civilisations on Earth; possibly offered to them by others. The hypothesis that the construction of the many megaliths thousands of years ago was achieved by manual labour alone, is preposterous. In spite of this impossibility, many so-called experts try to explain that the manipulation of stone slabs, some weighing 50 tonnes or more, was achieved by slaves using ropes and planks. Even present-day engineers state that using modern engineering machinery they would find moving such megaliths around pretty well impossible.

It is a frequently noted occurrence that cars in the close vicinity of UFOs lose all power and will normally come to a stop. This may be a deliberate action by the craft occupants but the random nature of the occurrence suggests the unintentional side-effect of an intense electric field extending outward from the craft that affects a car's electrics. Interestingly cars with diesel engines do not seem to be stopped, probably because these engines do not require an electrical spark in order to operate. This supports the hypothesis that gravity modification is mediated by an intense electromagnetic field.

Bob Lazar, famous for being an engineer allegedly working on alien technology inside Area 51, claimed that the element 115 (moscovium) underlies antigravity technology, but this is probably deliberate misdirection. Much more likely though, is that the physics of electrogravitics is responsible for the theory behind antigravity propulsion, as outlined above.

Humans, Humanoids And Brains

If the circumference of a Grey ET's cranium, is as estimated, about 10 cm greater than ours then it would increase their brain volume to around 2400 cc which is almost twice the size of our average brain volume of 1300 cc. The brain is an organ of finite size so it must also have a finite storage and processing capacity for information, rather like the hard drive on a personal computer. It may be necessary to have a larger brain volume if a life span is much longer and that extra information and processing power will require more storage space. If an advanced race many thousands of years ahead of us think that genetically engineering a larger brain is a good idea, then they probably know what they are doing. There have been a number of ancient skulls unearthed in the Middle East, North America and Peru which show abnormal shape and size; those from Peru are known as the Paracas skulls. They were examined by medical experts who stated that their size was not due to disease (hydrocephaly) or the tribal practice of deliberate cranial deformation. Mitochondrial DNA was extracted from the remains of the Mayan, Peruvian skull (a child) which was also carbon dated and showed that the mother of this individual was a human from around 1300 AD. The local legend in Peru was that earlier generations of the native Maya were visited by people from the stars and stone inscriptions of these visitors showed they had large heads. The mitochondrial DNA (inherited only from the mother) indicates the mothers were human but contain some previously unknown mutations. The nuclear chromosomal DNA was partially degraded so was harder to sequence and as yet no official results have been published. Of particular interest would be Y chromosomal DNA which originates from the father only. DNA extracted from the bones of these skulls or other material, such as hair, should be sequenced by several reputable laboratories to eliminate criticism from orthodox science which always struggles with the concept of ET visitations. In fact, the traditional interpretation of humanity's relatively recent past may have to be largely rewritten in the light of this type of hard forensic historical evidence. Unfortunately, orthodox science is very resistant and has its head firmly in the sand when it comes to considering any such new possibilities. The Paracas skulls' considerable elongation has significantly increased their cranial volume by 30% compared with modern humans. The possibility of such deformation being produced by binding the skull of infants can been ruled out as this procedure *cannot increase* the cranial volume. From a mathematical perspective, using integral calculus, it can be shown that the geometric shape of a sphere will always produce a maximum volume for a minimum surface area, so any artificial deformation or compression of a spherical skull after birth, will by definition, *reduce* the volume of a brain contained inside. Or to put it another way, you cannot expand the volume of a balloon by applying pressure to the outside of it. However, the elongated skull as a genetically engineered alteration may have enabled this race not only to have a much greater cranial capacity but also permitted normal childbirth. A larger spherical skull shape would have difficulty passing through the birth canal. Also, the position of the foramen, which is that point at the top of the spinal column where the skull is articulated, is in a completely different location from that seen in normal humans. This indicates that a process of skeletal development producing this alteration must have begun during foetal growth, rather than after the birth. The genetically determined repositioning of the foramen was necessary to allow a heavier, elongated skull to be balanced correctly on top of the spinal column. The elongated skull shape allowing a relatively normal, but probably difficult childbirth, differs from the more globular, spherical skulls observed in many ETs like the Greys, who must be born from artificial wombs. The very similar shape and size of such elongated skulls in widely separated parts of the world suggests a common source that may reflect an extraterrestrial origin. Drawings of Queen Nefertiti from ancient Egypt are also depicted with an elongated skull and some skeletons excavated from Egypt from that era show this unusual cranial formation. But Egypt and Peru are 7000 miles apart and they both have historically, civilizations rather fond of enormous pyramids. It is possible that a particular race of ETs, the Annunaki discussed below, may have visited several ancient Earth civilizations and interbred with the locals, possibly attempting to jump start our development more than 5000 years ago. The TV series *Ancient Aliens* feature presenter and author Giorgio Tsoukalos with his team covering this subject well, with film-footage of the ancient skulls. The presenters of this series speculate that the elongated skulls may have originated from a race of more advanced beings who occupied Atlantis before its destruction. A small number of them managed to escape and were scattered to various places around the Earth. Of course, it is not the absolute brain size that is important but the relative brain size to body mass. The whale brain is around four times larger than ours but the ratio of brain size to their body mass is much smaller, as their bodies can weigh over 100 tons. The brains of Neanderthal man were around 1500 cc but they had an estimated body mass much greater than modern man. But there comes a point that if human brain volume is progressively reduced, eg. in conditions like congenital microcephaly, then a very definite reduction in intellectual function is seen. The opposite would be the case with the little Grey ETs, as not only are their brains larger than ours but an estimate of their body mass must be around half that of a human. They are reported to have a body size similar to that of a human child. I have portrayed a diagram below of the significant hominid skulls throughout the last 6 million years of evolution that includes the more unconventional skulls like those from Paracas and the Yeti (Bigfoot or Sasquatch). Officially there are no Yeti remains but artistic license has created a probable skull likeness and we could speculate that these primates would have a cranial capacity of around 900 cc.

There are some concerns about the direction of human intelligence. Firstly, examination of ancient human skulls shows that there has been a slight reduction in average cranial volume of about 10% in the last 12,000 years according to M. Henneberg (33) and there is some correlation between general intelligence (IQ) and brain size. Secondly, although there are probably no hard statistics on this, a glance around the planet suggests that people of below average IQ tend to have more children than those of above average IQ. As long as people continue to have children, then evolution never stops and the direction of evolutionary pressure will therefore gradually drive down the average IQ of the human race. Studies on human twins have indicated that the hereditary component of intelligence is about 75%. No doubt academic psychologists and epidemiologists would be fairly sniffy about this sort of conclusion and say it does not matter. However, if humanity is getting stupider this is not a trend to be welcomed. Also, the so-called Flynn effect that saw an apparent increase in IQs during the 20th century has now stopped and even reversed in Western countries (34). There may come a time in the not-too-distant future when medical genetics has advanced sufficiently and our ethics have become more realistic, so that our intelligence can be rescued by genetic modifications to the human brain. A recent scientific assessment of intelligence and brain volume was led by Jansen, P. R. in *Nature Communications*, 2020 (35).

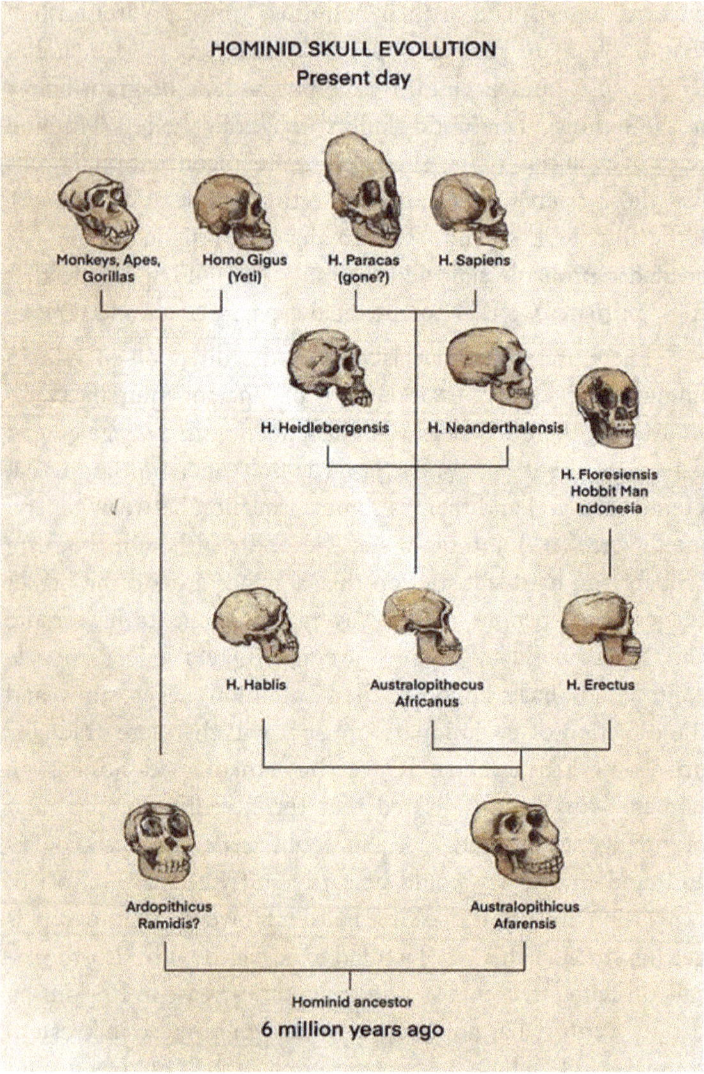

THE 4-DIMENSIONAL AFTERLIFE AND OUR PLANETARY CHANGES

Figure 4. Skull Evolution

Disclosure

We can be almost certain that some ETs and obscure groups in the defense and security hierarchy of the US have established a form of working relationship which must have happened sometime in the years after World War 2. However, as recently as 2024 a Pentagon report stated there was no evidence that the US government had covered up the existence of aliens. Earlier in 2010 there was an official US government denial of ETs when Phil Larson a spokesperson for the President Obama administration and NASA said: "there was no evidence that any life exists outside our planet or that an extraterrestrial presence has contacted or engaged any member of the human race". This statement was either amazingly ill-informed or dishonest. We do not know as yet what exactly ET is doing here but it will be connected to the profound changes that will shortly take place on our planet and some preparation for these events is necessary. The last thing ET wants is trigger-happy US Airforce pilots constantly taking pot-shots at them. ET could drop our jets out of the sky at the touch of a button if they wanted to, but they are here to set an example of peace and future co-operation so confrontational situations must be avoided. Exactly who the black-ops groups involved with ET interaction were, is above top secret, but probably involved highly exclusive sections of the CIA, the National Security Agency and the Defense Intelligence Agency. These are often referred to as Unacknowledged Special Access Projects (USAPs) and members would be recruited on a need-to-know basis. There is speculation that a privileged group within NASA is also aware, it would be difficult for them not to be, as they regularly spot mysterious craft buzzing around near-Earth space. More recent US presidents are unlikely to be in the know, especially if they are Democrat as they can't be trusted. President Truman initiated this post-war secrecy with the so-called Majestic 12, a highly placed group of individuals mostly from defense and security. President Eisenhower allegedly had a meeting with a group of Nordic-looking ETs to establish a form of cooperation in 1954. The US relationship with ET must have been running for decades now and may also involve other countries where there would be some arrangements on secrecy. Otherwise, the US could be placed in a slightly awkward position if one of them should get in first and grandly announce that humans have made contact with ET. The location of such interactive bases is no big secret, eg. Areas 51 and 52 in Nevada, The Dugway Proving Ground, Utah, the Wright Patterson Airforce Base in Ohio and many others. The intense levels of security in these bases indicate a fantastically high level of official paranoia way beyond the concerns about say, someone simply taking a snapshot of a new stealth fighter. They really do seem to have something to hide.

There is an interesting possibility here that the US military is most certainly aware of but has gone very quiet about and that is extra-sensory remote viewing. Many of the US security agencies in the 1950s and 60s realised that a few exceptional individuals were capable of psychically viewing distant geographical locations remotely and their ability could be improved by training. Not surprisingly, they were concerned the Soviet Union had people who could do the same. But a few years later it all went quiet as the US security and defense establishments officially dismissed remote viewing as a waste of time and money. However, this needs to be translated as: we have this technique which at times is amazingly successful and we don't want other people doing it. This ability is not quite the same as an OoB experience as it seems a person's astral body is not detached from their physical one during this process. Obviously hostile powers having visual access to US military installations would be a big problem. An ex-remote viewer from the US military, Jim Schnabel has written a very interesting account of what went on in *Remote Viewers: The Secret History of America's Psychic Spies* (36). The extreme levels of paranoid security at locations like Area 51 could be effectively rendered worthless if multiple individuals with this remote viewing ability were, to corroborate independently, on what they could "see" inside the buildings and tunnels of this constantly expanding military base. If they all saw the same thing, then their collective conclusion on what was there would give the accuracy of such viewings a very high level of credibility. There are a number of private remote-viewing organisations in the US who should be encouraged to look inside these hidden bases, that is if they are not doing so already. A quick check on the internet will show that some individuals are attempting to view these places but the conclusions they reach are typically rather lurid and not shared by others. Trying to observe Area 51 and many similar secure bases from ground level is a hopeless task given the high levels of security.

The decision to keep the matter of human and ET contact secret was made at the height of the cold war with the Soviet Union, so of course a difficult time. Unfortunately, those individuals who reach high places in the military command and in security agencies are not often distinguished by great intellect or vision and they will have found themselves in the awkward position of having to make the biggest judgement call in the history of humanity. And what a surprise, they failed it. In particular, they were also able to silence those presidents who leant towards going public. The pressure for such secrecy around this matter seems quite extraordinary, even to the point of authorizing deniable killings. Paranoid, myopic military and security service leaders do not want to change the status quo, they saw panic in the streets, civil disturbance and people hiding under their beds with tinfoil around their heads. Of course, the people cannot be trusted and they must be protected from themselves. The less we tell the people the better. Denying the truths and reality of our universe, by a small group with power and control who have deluded themselves into believing that they know what is best for us, is a disgrace. Sadly, this is what we get now from a patronising elite who manage to reach high places and take a very backward-looking analysis. Quite the opposite will happen: people will be dancing in the streets asking why has it taken ETs so long? There will be great relief that there is a higher source of human (humanoid) wisdom and experience in the cosmos that we will soon have access to. In fact, those people responsible for blocking disclosure should be furiously polishing up their excuses, because when disclosure does occur, they will be first up against the wall in any revolution that follows. There seems to be, in the US and probably elsewhere, a parallel government, a hidden cabal, deep within the state that oversees these matters and keeps elected leaders out of the loop. The suppression of this information regarding the existence of ETs and the level of disdain shown by those in the know, for the rest of the world's population, is essentially criminal. Especially when we consider the incredible technologies that ETs would make available to mankind on a planet badly affected by shortages of resources and energy. ET's contribution to an understanding of humanity's health burden and possible treatments, especially with regard to the aging process would be profound, as their healing concepts take place on a metaphysical level. The decision to withhold all information regarding the existence of ETs and the obvious technological contribution such advanced civilizations could make to solving so many of mankind's problems, indicates an extraordinary absence of ethical concern for humanity. The greed for an exclusive hi-tech military advantage has vastly outweighed any humanitarian considerations to put the needs of our planet first. There is only one overarching priority, and that is; we must have military superiority and the extraordinary massive allocation of funding that it demands. So, God help us. An in-depth speculation on the consequences of our contact with ET is explored fully in *A.D. After Disclosure* by Dolan and Zabel (37) and there are many enlightened accounts of why disclosure has been blocked by Dr Steven Greer (38). NASA has known very well for some time that we are being visited by ETs. It will be interesting to see how they deal with the private explorations of space now being pursued by multi-billionaires like Elon Musk, Jeff Bezos and Richard Branson. If they succeed soon, in landing on, or orbiting the moon, they will start seeing all the stuff NASA has been trying to keep secret from us for several decades. On numerous occasions the live video feed from the Apollo missions and the international space station have been cut when anomalous phenomena have been spotted by the ground staff or astronaut crews, with rather implausible excuses. (As some wags have pointed out NASA should stand for Never A Straight Answer). However, NASA is a very multi-layered organization and probably only a few well-chosen individuals are on the inside team. How will these private space entrepreneurs and their teams be silenced when they also spot these mysterious structures on the moon and inexplicable craft? Maybe they won't be and the cat will be out of the bag. Possibly those who have been keeping the UFO question above top secret are hoping that disclosure will take place gradually and subtly, so they will not have to provide explanations for their paranoid secrecy beginning in the late 1940s that prevented the world from being given the most important cosmic information affecting Mankind forever.

ETs could have bought off US cooperation with the promise of some technical goodies. With technology many thousands of years ahead of ours they have much to offer. The prize piece of advanced technology is of course antigravity propulsion. This technology will revolutionise transport, replacing any form of propulsion requiring a flame like internal combustion, jets and rockets and supersede the electric motor. However, giving US scientists a small UFO and asking them to back-engineer it might be a tall order without a lot of help. It would be like going back to the Victorian age with a smartphone and asking them to make another, which would have been almost impossible for them with their current technology. Also, there would have to be some conditions for handing over antigravity devices, especially with regard to forbidding their use for military purposes. If the US secret science programmes have been able to reproduce antigravity technology with ET assistance, then they must be already travelling around our solar system within days rather than months or years. Of course, the public would be completely unaware of this. There have been a number of reports of supposed UFO crashes, in particular the well-known occurrence at Roswell, New Mexico in the USA in 1947. Events like these may have obliged ETs to come to some arrangements with highly placed military commanders and security leaders in the US. There are claims from some sources like Dr Steven Greer's Sirius organization (discussed below) that most of the research attempting to reverse engineer ET technology has been outsourced to unacknowledged private contractors in the aerospace industry. This ensures that even people highly placed in the government and defense departments can be kept out of the information loop with plausible deniability. The scientists working on these projects have been referred to as the "invisible college". The private undercover organisations and their political masters trying to prevent disclosure are sometimes referred to as the *Silence Group*. I am a very reluctant supporter of conspiracy theories but I think there is definitely one lurking here.

Another rather unfortunate possibility is that the US is taking a very mercenary view of ET cooperation. If the most amazing area of technology available from ETs is antigravity propulsion, then after disclosure, the rest of the world will badly want access to this science. Maybe ETs will provide, but we cannot be sure so possibly the US has manufactured large numbers of smallish antigravity craft and is storing them in one of their enormous underground facilities. These machines will be available at price and whoever is manufacturing them will make billions. So, the US will eventually reveal the truth about what they know of ETs but only on their terms, probably a bit risky. I hope this will not happen because the New Age of spiritual enlightenment and cooperation which is expected may usher in a very generous world where personal greed, profiteering and even money will become redundant. In our universe there is an abundance of free energy but unfortunately our restrictive, orthodox science has not yet worked this out, or it has been suppressed. There is currently much alarm about the possibility of future global warming and climate experts say man's use of fossil fuels is responsible due to the production of CO_2. The novel sources of energy that ETs are utilizing to power their spacecraft are enormous, possibly extracted from the so called zero-point energy that some theoretical scientists suspect must exist in the vacuum of space. But currently we do not have the technology to gain access to it. This huge abundance of unlimited energy would free our world entirely from fossil fuels and the highly unreliable sources of renewable energy like solar and wind. Yet in spite of our planet's predicted climate danger, we have a small group of privileged people in high places (mainly the US?) who are hiding the knowledge about ETs with their revolutionary technology and are refusing to tell the world. This conspiracy of secrecy shows a contempt for all of mankind. Those privileged USAP groups within the US military who have access to antigravity propulsion and other advanced technology understand that it cannot be used overtly for military advantage for two probable reasons. Firstly, there is an arrangement with ETs that gifted technology must not be used for the purpose of warfare. Secondly, if there were any military confrontation with terrestrial adversaries then the use of such weapons with staggeringly futuristic science will immediately give the game away that the US military has acquired this technology and the whole world should really know about it and is entitled to it. So, I suspect the power this technology brings is being used more subtly, especially if it is off-world. For example: military satellites, any space probes to the moon or Mars launched by states regarded as a threat, would be considered fair targets for sabotage and there would be complete deniability of any involvement by the US. Preventing the deep-space probes sent from other nations to view the dark side of the Moon or obtaining close-ups of Mars would be very much in the interest of the US government and USAPs because they would of course spot the structures that NASA does not want the rest of us to see. Many Russian probes to Mars have mysteriously disappeared or malfunctioned in the last few decades. A recent example of note is the failure of Russia's Moon-probe Luna-25 in August 2023. The Roscomas official explanation was that the device experienced an "abnormal situation and switched to an unplanned orbit" then crashed. Well possibly. But if American USAPs have the technology that gets you to the Moon in minutes, then giving such a probe a slight bump will crash it and who would know? There is an incidental bonus here in that one of President Putin's prestige science programmes is delivered a serious blow.

However, there is an alternative scenario. ETs could be pulling the strings. ETs will be concerned that antigravity propulsion (space- time warping technology or trans-dimensional travel short-cutting through the lower astral planes) will enable people from Earth to travel to other solar systems with very short journey times as the speed of light limitation will not apply. They do not think we are quite ready for this. We are typically a belligerent race, spiritually immature and should not be inflicted on other inhabited planets. In fact, they may be actively trying to prevent scientists here developing the technology that will enable interstellar travel, until we are considered to be ready. The two most important developments in the history of mankind that have propelled civilization forward were, firstly the discovery and control of fire and secondly, the discovery and use of electricity/magnetism. The third will be gravity manipulation. However, it seems almost like this last breakthrough is being actively withheld. Prodigies in the arts and sciences who may appear gifted, have often been given subliminal and subtle help by our fellow beings in the metaphysical dimension. So, from time to time, they are able to nudge us along in the right direction by stimulating creativity and understanding. But some areas in technology may be blocked for our own good. Alternatively, we could give the independent Deep State, hidden government in the US the benefit of the doubt. They may well be anticipating what happens if the technologies of free unlimited energy and antigravity propulsion are released to the world, as there would be many unintended consequences. Of course, the US world dominance with regard to military power, control and energy provision would quickly disappear. Whether we like it or not, the US with a history of settled democracy does provide a certain amount of global stabilising influence. The problem starts when rogue states, dodgy dictatorships and places with unaccountable government get their hands on this technology. If the unhinged leaders of some of these states had the technology to travel anywhere in the Solar System and maybe beyond, the consequences are frightening. The problem is obvious: an unenlightened humanity on this planet is not ready.

ETs of course, could resolve the question of their existence, literally overnight, if they wished. A worldwide, non-threatening, display of their spacecraft at strategic locations would allow millions of people to see them and the 24-hour news media would do the rest. This is what some call catastrophic disclosure. In all probability, various ET groups will have been in this situation before, with other emerging planets about to make the next leap in cosmic and spiritual awareness, so they will have a plan of action. ETs might then go away for a while after revealing themselves; maybe for months or years so that humanity can digest the implications of what they had seen and prepare themselves for a more inclusive and peaceful interaction in the future developments. This is the classic sci-fi theme with ETs landing on the White House lawn. But they have chosen not to do this and we have to ask why. Or rather more worryingly the question is: what are they waiting for? In fact, ETs are generally still attempting to be fairly secretive. ETs are likely to be unhappy with the conventional politics on planet Earth; they would prefer to see a global awareness with a people-led acceptance of their existence, rather than some narrow, state-controlled announcements from venal governments. Their policy seems to be: we will give you glimpses of our existence to arouse your curiosity, but not too much. They seem to be deliberately behaving in a way that is confusing and ambiguous, so we are not able to understand what is going on. We have to examine the reasons why ETs do not wish to disclose their presence to the wider world at this time. Another point of interest is that allegedly the Greys have been producing hybrids between their race and humans on Earth, an issue which has distressed some female abductees. The idea being that when universal contact does take place, then these hybrids will be able to bridge much of the gap between our two races. As yet, these hybrids are not yet mature enough to have much of an intercessionary role so presumably when they are older the time will be right.

One hypothetical possibility is there may be a Rapture-like scenario coming fairly soon. Some fundamentalist Christians believe that in the "end-times" the chosen few will be saved from the grim events of the Apocalypse by being somehow spirited away. This could be a type of supernatural phenomenon beyond our current understanding. Alternatively, the Chosen People could be lifted from a devastated planet Earth in large UFOs. ETs may already be selecting the Chosen People by using some telepathic or spiritual criteria based on personal cosmic awareness and spiritual conversion; or possibly the specific individuals chosen may have had previous incarnations on other planets so already have close bonds with ETs. Some individuals claiming alien abduction believe that small implants have been placed inside their body, which would enable ETs to locate them quickly if necessary, when there is some degree of urgency. A number of sources like the French prophesier Nostradamus and the Bible predict apocalyptic events in the "end-times" which will include the appearance of an Antichrist followed by the return of Jesus who defeats the Antichrist and brings forth a heaven on Earth. These times cannot be far off. However, many ETs are extremely psychic with superior visualising paranormal abilities like remote-viewing or precognition and they must be fully aware of these impending disasters. Furthermore, if an Antichrist is coming soon, it is quite likely he is already living here on Earth. So why do ETs not tell us who he is and where he is? Perhaps they are waiting for a world government that would be formed after some apocalyptic event and they can deal with humanity as one.

Another contender for the Antichrist is a corrupt world government that has come into being under the false pretense of global unification and saving this planet from a threat. The message that the world will be deceived in the so-called end-times is made clear in the Bible. The *Sirius Disclosure* organization (discussed below) believes that a hidden, alternative government is already operating in the United States and has tentacles of power reaching transnationally; it could be described as a form of breakaway civilisation. This totalitarian and militaristic hidden government has access to exotic technologies and deep black projects, some of which are reversed-engineered extraterrestrial and currently not revealed openly to the world. If they were, this advanced technology would seem miraculous to ordinary people when revealed to the world during an existential planetary crisis. The shadowy forces behind this plan want ultimately, military power and global control; they are ego-centred not God/cosmic conscious-centred, in effect an Antichrist in the waiting. Again, a theme high-lighted in the Bible's Revelations.

Alternatively, ETs may well share the views of those obscure and unacknowledged security overlords that disclosure will be too disruptive for humanity to handle, when all our religious, scientific and historical beliefs are overturned. But more likely, it is terrestrial vested interests who want to protect the status quo because they are doing very well from it. But the enormous revelations on this subject will have to happen, sooner or later. The hierarchies in many religions will have to deal with the fact that Earth's humanity is not special and their assumed power and authority will vanish. There will be questions like: do all planets have a Jesus? Is our son of God unique? How many other humans/humanoids are in the universe? Reincarnation will be another thorny issue when confirmed by ETs because people will have to understand that their circumstances here on Earth may be due to what they did in previous lives and there is little point trying to find someone else to blame. Many will be reluctant to accept this, although many millions of Hindus do so already. There have been occasions when people with a high media profile have pointed out that the unfortunate life circumstances that some of us are in is likely due to our actions in earlier incarnations. In the current, fabulously judgmental social media age, we find such outspoken people are heaped with abuse or "cancelled". It is no longer acceptable to say that individuals are responsible for their present life situation and we must find someone else to blame. However, when the new age of a greater enlightenment and awareness arrives the connection between our present circumstances and what we have done before will become obvious. Having a higher source of knowledge and an awareness confirming that death is not the end will hopefully be a great comfort to most of us.

The Sirius Disclosure Project

This is an organisation based in the US dedicated to revealing the truth about ETs visiting the Earth. It is headed by Dr Steven Greer and they are doing an excellent job attempting to bring this information to the world. This group has realized that for a long time now the shameful official policy of the US and probably other nations, is to deny all information, regarding the presence of ETs, to the public. So obviously non-governmental organisations should step in and lead the way and so they must. Dr Greer and his team have put themselves under a considerable amount of personal risk as the shadowy groups attempting to prevent disclosure or the release of exotic technology are not averse to killing in order to maintain secrecy. However, Dr Greer and his team do risk damaging the overall thrust of this message to the world, by presenting a number of outlandish claims which have questionable plausibility and are more akin to conspiracy theories. Of course, the big problem with these claims is that there is always a serious shortage of real evidence and we wait desperately for hard data. One of the most suspect assertions from this organization is that ETs are always embracingly benevolent and would never engage in the abduction of humans. This must sound very surprising to the huge number of people who claim to have had this experience, especially in the US, and are often associated with extraordinary paranormal circumstances. The Sirius Disclosure group believes these abductions are staged by covert US Black-Ops groups and the "aliens" are a type of manufactured nuts-and-bolts robot assembled in unacknowledged laboratories in the US. Maybe Dr Greer should discuss this assertion with the likes of Whitley Streiber who has written several books on his abduction experiences. Streiber experienced the full range of psychic and paranormal interactions with the little-Grey ET beings, including: 2-way telepathy, levitation, OoB travel and the projection of mental "films" directly into his mind. There is not even the remotest chance that any of the 3 letter security agencies in the US, or their out-sourced labs with allegedly captured ET technology, could have a robot with these capabilities in the 1970s or 80s, and may never have it for hundreds of years. Another dubious claim is that these Deep-State organisations within the US will attempt to justify an attack on the ETs claiming they are a threat to planet Earth and humanity and hope to bring about an "End-Times" Armageddon event described in the Bible's book of Revelations. These organisations have managed to reverse- engineer and weaponize some ET advanced technology and intend to justify the use of it against the ETs after a false-flag attack on Earth. There are some problems inherent in this assertion. Firstly, the ET civilisations are, in most cases, many thousands of years ahead of our own. They have been watching us for eons and will have an awareness of what is going on down here and are certainly not going to be duped by malign covert groups in the US attempting to set up a false-flag operation implicating ETs as hostile. ETs understand that the vast bulk of humanity here, is by their standards, unenlightened and uninformed but essentially innocent; I do not wish to use the word ignorant. Even with some acquired ET technology, attempting a war with ETs is laughable; we would have no chance. The advanced ETs can telepathically get inside the head of any soldier or pilot attempting to target them and mentally "sedate" them, so forget about death-rays and kinetic weapons being of much use. Dr Greer also fails to adequately explain why the ETs in most close encounters with humans go out of their way to remain elusive and secretive; this contradicts his claim that ETs are open to contact and disclosure. In fact, normally ETs attempt to psychically block the contact experience and many claiming abduction resort to hypnosis in an attempt to find out what happened to them. Finally, those individuals belonging to the covert organisations attempting research into areas of reversed ET technology like trans-dimensional experimentation, are human beings. They will have become very aware, rather more than most of us I suspect, that a non-physical afterlife dimension exists and this will include a place for themselves. They must know, that one day when they lose their physical bodies, they will end up in one of these places. Obviously, it would be very desirable that their personal destination in the metaphysical astral dimensions is not too uncomfortable. Actions have consequences and at some point in their future they will be confronted by their higher self and a life review process where explanations will be required about their life in the physical plane and what they achieved this time around. There may be serious karmic implications for those individuals who have been accomplices to murder and willfully blocked progress that will bring such incredible benefits for the common good and humanity. Greed and selfishness are not rewarded in the higher astral planes, especially if you knowingly did not cooperate with the divine purpose that embraces our universe.

Dr Greer of the Sirius Organisation has pointed out a number of times that the advanced interstellar ETs with the large heads have average IQs about 5 times higher than ours. This begs the question: then what chance do we have? We are supposedly a level-zero planetary civilization in galactic terms that must shortly move forward. How can we reach the higher levels if our intellects are so relatively backward?

Technical people with the Sirius Disclosure group are also attempting to construct on an electromagnetic and mechanical level, various types of free-energy devices that are currently being blocked from world use. They would offer this technology on a free open-source basis to the world and let us hope they succeed. Unfortunately, a few private individuals or small companies who have stumbled upon this technology inevitably attempt to monetise and patent their discovery. This is a big mistake, as the unacknowledged, hidden powers protecting the status quo either quickly pay them off or issue threats to eliminate them, possibly by unpleasant means. It is disappointing that after many decades of rather glib claims, especially in the light of many YouTube clips about such free energy devices, nothing has been revealed. I suspect that the spectacular lack of hard evidence as to their workability means the technology involved with harnessing zero- point energy is much more difficult.

The access to unlimited zero-point energy has such profound implications that we can see why a clamp-down on this information has been put into operation by the dark forces of clandestine governments, probably operating transnationally. Firstly, access to this free energy will mean the multi-trillion petro- dollar power generation industry based on fossil fuels and eco- friendly sources like wind and solar will become obsolescent almost overnight. Secondly, and equally important, are the many military-industrial complexes around the world that have relied on the manufacture of kinetic technologies for weapons systems and defense. As soon as, almost unlimited, electromagnetic energy becomes available, then warfare will exclusively employ energy beam weapons of some description. Laser weapons are now at the stage where drones and small planes can be brought down within seconds by a laser beam at a tiny cost, for example the Dragonfire Laser System soon to be deployed by the British navy. These lasers are in the kilowatt range. Once huge energy sources are instantly on-tap then megawatt lasers will be available and this will be a complete game-changer from the military perspective. All kinetic weapon technology will become effectively redundant. Planes, missiles, rockets, tanks and even warships will have no defense against a high energy-beam weapon that can direct a huge amount of destructive energy within seconds. Do these two reasons justify blocking our civilisation's access to a free, unlimited, non-polluting energy source?

On the few occasions where there have been close encounters with communicative ETs without psychic sedation they make a point of admonishing us for the way we damage our environment and the planet generally, mainly because our power generating systems are relatively primitive. It would not be that difficult for them to pass over some details on where to start with extracting zero-point energy to a significant number of people. If this technology will save our planet then why not do it?

Ancient Aliens And Our Evolution

Conventional science is quite unwilling to countenance the possibility of human/ET hybrids or even the possibility of other human sub-species. For example, it took many years for mainstream science to accept the existence of "hobbit man" or Homo Floresiensis discovered on the islands of Indonesia in 2004. The discovery of this species that existed around 17000 years ago was initially resisted and ridiculed. The local legend in Peru has it that earlier generations of the native Maya were visited by people from the stars and there are stone inscriptions and pictures of these visitors showing unusual faces and heads. A few small studies of unusual ancient skulls from several different worldwide locations have been undertaken by reputable researchers in their own fields, so far with indeterminate results, as Y chromosomal DNA (male) was not obtainable. But mainstream science and medicine has studiously ignored these investigations and conclusions. The conservative nature of scientific research is quite a barrier to the further exploration of extraordinary discoveries which do not fit into orthodox views.

This leads us to another particularly awkward possibility regarding our heritage that may upset many. It is very likely that some ETs have influenced human evolution in the distant past, or even interbred with us at various points in our evolutionary history. How we deal with this information may be difficult but I think the evidence is overwhelming. Any intervention in human evolution by extraterrestrial humanoids would be an extraordinarily long- term plan. Way beyond the life spans of even long-lived ETs, so any master plan must be approved on a grand scale with continuity over many generations. If the process of human evolution had stalled about 1 million years ago then the higher cosmic and spiritual authorities may have decided to give things a bit of a jump-start. This would have required some tinkering with the DNA of Earth's most advanced primates in order to speed things up, with assistance from advanced ETs. This might explain why the so-called "missing link" in fossil evidence has never been found, because there isn't one. The genetic evidence shows that as a race, humanity here on Earth arose from this planet. For example, our genetic material DNA has a 98% match with chimpanzees and there is a gradual reduction in DNA homology from man relative to other species the more distant they are from us. We even share many of our genes with the simple unicellular organisms like yeast. This is strong evidence, therefore, that we belong to this planet and are related to the rest of life here. If ETs had intervened in human evolution, it could have been when the brain of developing primitive hominids trebled in volume from around 400cc to 1300cc and that was about ½ million to 1 million years ago. This was when, we believe, that pre-human species like Homo Habilis and Homo Erectus were evolving towards Homo Sapiens. The first fossilised evidence of Homo Sapiens skulls is from around 300,000 years ago and was published in *Nature* (39). The population of developing higher primates at this time would have been fairly small so introducing modifications to genes that are responsible for increasing cranial capacity would have been quite feasible. Once the brain has reached an acceptable minimum size then the forces of evolution can proceed according to the natural conditions of the planet. Of course, this is speculation and it may well be that rapid evolution alone could account for such an increase in brain size. A further possibility is that early man was deliberately brought to Earth from other planets. A number of

ETs witnessed in close encounters have been indistinguishable from us here on Earth. In this case a common genetic origin is highly probable, but from where?

A further intervention may have been around 10 to 15 thousand years ago, shortly after the end of the last ice-age. Specific ET races were given "permission" to push the stalled human progress along, again with the oversight of more spiritually advanced beings; this would all be part of God's purpose to facilitate the development of human civilization throughout the cosmos. The race involved may have looked fairly human but not completely and there are many legends of god-like races interacting with us in ancient history, in particular a race known as the Annunaki. The story has it that they were sent here to benefit humanity as much as possible but things went wrong. They had human weaknesses and began to enjoy the status of being gods. They of course brought with them the advanced science of a much older humanoid civilization. This race was a likely source of the unusual skulls and giant skeletons often found near megalithic structures around the world. They brought with them advanced technologies like interstellar travel, incredible energy sources and antigravity devices that were used to build the megalithic structures around our planet. Of course, when they left these machines were taken back with them. Many of the megaliths, like Stonehenge, that still exist were built in such a way to enable primitive man to predict the movements of the sun, moon and the solstices. This information would be absolutely essential for any civilization to progress as it will enable farming and agriculture that rely on predicting the seasons to be fully productive. Very often archaeologists have marvelled at how ancient races like the Mayans and Egyptians had developed so much advanced astronomical and mathematical knowledge, well of course they didn't: it was given to them by the alien visitors.

When the Annunaki arrived initially they must have set up their base in the region that is now Egypt. The legends and records in Egyptian hieroglyphics indicate they came from a solar system in the constellation of Canis Major near the star Sirius or the dog star. There has been much speculation as to whether they bred with humans and this was very probable as they were human enough with a good genetic match for hybridization to occur. We may all have traces of Annunaki DNA in our chromosomes, like we do with Neanderthal DNA, estimated to be around 2%. The human/Annunaki hybrids were called the Nephilim and are the most likely candidates for the origin of Paracas skulls that show extreme elongation with increased cranial volume. The Nephilim were described by The Bible in *Genesis 6* as a race of giants who existed before The Flood in the Eastern Mediterranean and could be a source of the legend involving David and Goliath. They were not a viable race long term, possibly for genetic reasons (hybrids often are not) or they were killed off because of their slightly unusual appearance. Maybe this was another attempt to increase human brain capacity on Earth and with it an intellectual improvement. We see again that there are higher cosmic forces who believe that humanity needs an intellectual upgrade and looking around our planet you can see why. The volume of Paracas skulls is about 1500-1700 cc or 25% greater than ours. An in-depth study, into the Paracas skulls, with photographs, was published by Brien Foerster in *Elongated Skulls of Peru and Bolivia: The Path of Viracocha* (39). Such interpretations of ET's intervention into mankind's long-term history are supported by Dolores Cannon in her conversations with hypnotically regressed subjects. But we have to be cautious about information from hypnotic regressions as it can be very prone to confabulation from subjects who always feel obliged to provide information.

Summary

As I write this in 2025 there are still no immediately obvious portents of the earthly events that will change humanity forever. But all the scholars and speculators in eschatology say it cannot be long now. The Bible says that the Lord will come like a thief in the night, so expect a surprise. We must hope that it will be a peaceful and exciting process, without widespread destruction. There have been many doomsday cults with soothsayers and gurus predicting the end of the world and announcing dates, sadly often with unfortunate consequences for their followers when mass suicide is encouraged. So far, they have always got the date wrong but on the basis of probability sooner or later, one of them is likely to get it right, however it will not be due to any particular gift of their prophecy. The highly publicised Mayan calendar predicted 2012 as the end of the world but not much happened. Researchers on the writings of Nostradamus are very divided about his predictions regarding the end-times. ET disclosure will likely hasten events even if it becomes a rather drawn-out and manipulated process. There have recently been some changes to US laws that protect whistleblowers. This has resulted in a number of ex-military men coming forward and they are attempting to convince American congressmen that covert scientific research projects have back-engineered alien technology and have the bodies of extraterrestrial humanoids hidden in underground bases. If hard evidence is provided about this then the world will change forever but exactly how is a bit unpredictable.

The physical 3-dimensional universe is a uniquely special part of God's creation that has been "zoned off", for want of a better description, for the benefit of humans/humanoids. It is a dimension where the rules of existence are different and more extreme (harder?) and it requires effort in order to create, take risks and survive. To use that rather corny expression; it is character building. In the physical dimension our body acts like a cocoon and blocks us from the multitude of thoughts and feelings of other living beings around us. So, we may not experience immediately the pain or the pleasure we have brought upon others while we are here. Very often in the physical realm we feel obliged to pick a side whenever there are differences in opinion. We must learn to accept that there are often just different ways to look at the issues that divide us but we must reject violence. People who have had OoB experiences recall that when in the OoB state, they are able to focus and become aware of a network of fine metaphysical filaments that connect them with every other being in the universe.

When we interact with others these filaments become more pronounced eg. with those in our same family or friends and work colleagues. Of course, the interaction can be positive or negative, but it is strengthened. If it is a negative link then there is disharmony that will continue to provide a tension until it is dissipated and it may continue into the astral planes after death and onto further Earthly reincarnations in extreme cases. This resolution requires the cooperation of both parties and the tension may prove to be disagreeable for the guilty party. Imagine what someone like a serial killer will experience when they arrive, presumably in a low astral plane, and are obliged to experience over and over again what they have done to their victims at a very personal level and from the victims' point of view. However, positive interpersonal links will strengthen and continue within soul groups who enjoy being or reincarnating together.

The occasional trans-dimensional manifestations and movements intruding into our physical world must seem to us, in our current state of knowledge, rather like magic. But once we comprehend that we exist in a multi-dimensional Creation then this must be acknowledged as a new field of science and potential technology. Of particular interest will be the many possible developments in the field of material science. The properties of metaphysical intrusions into the physical world are fully understood by advanced ETs and are used extensively as part of their technology, including some of their activities on Earth. According to the Sirius Disclosure organization some quite large solid objects for example, interstellar spacecraft, are mentally conceived in the metaphysical dimension and are somehow realized or extruded into the physical world. The properties of such solid objects when acquired and investigated by our scientists are baffling because on the atomic scale there is a level of isotopic composition and arrangement never observed in normal elemental matter. The concept of realizing metaphysical objects into the physical dimension is currently a very rare phenomenon or a mystical event on our planet, for example like those recorded in religious texts, including the Bible. Hopefully in the coming new age of enlightenment we will understand and acquire this technology so that mining, plundering the Earth's resources and metal-bashing factories will no longer be necessary.

The vast majority of the human population on Earth is fairly non- psychic and many animals especially our pets also seem to have some basic psychic awareness, occasionally better than our own.

Most of us have the occasional experiences of strange coincidence, moments of inexplicable intuition and fleeting glimpses of people on the astral levels, but none of the big stuff like continuous two- way telepathy, voluntary levitation and a clear foreknowledge of future possibilities. Anything like this would come under the current definition of extreme paranormal. Such psychic abilities are a reflection of our metaphysical heritage in the astral regions and obviously at some point we all return there. In the astral regions these abilities are normal and fundamental to existence there and beyond, where we no longer have physical bodies with all the imposed limitations they carry.

Some advanced OoB experiencers and shamans describe the Universe as a holistic and conscious unity. Mainstream scientists in the physical sciences will, no doubt, view this as ridiculous. All the information in the Universe is available to us individually, if only we were able to access it. We could compare this situation to a fragmented holographic image. When a holographic image is broken into many pieces then each fragment will still contain the original image but with a reduced resolution. Each one of us individually should be aware that we are all one of those fragments and this universal knowledge is wrapped inside the subconscious mind of everyone. This information is currently hidden within most of us but should change when our planet is transformed to a higher level of consciousness. The concept of the Universe being conscious will seem a bit baffling to most of us. But it goes further than that, as OoB experiencers and researchers like Rupert Sheldrake would claim that simpler life forms and even inanimate objects seem to possess this consciousness. This is not a consciousness pertaining to intellect or thinking but rather an awareness. So even a rock has an awareness of the emotional state of a person in the vicinity around it. The possibility that simple non-sentient life forms like plants may have this awareness has been explored by a number of researchers over the years. For example, the scientist Cleve Backster in 1966 wired up plants to a basic galvanometer which measures electric current, it is a bit like a lie-detector. When an intention by the experimenter towards the plant was made, either to nurture it or to harm it then a response was detected on the meter. It is difficult to explain how a life form without a nervous system can respond from a distance to human emotion. This so-called Backster effect was investigated more fully by Tompkins and Bird in *The Secret Life of Plants* (41).

We live in a holistic, conscious universe where the supermind of our Creator (God) pervades this creation in a way that we currently do not normally fully appreciate or understand. But at the central focal point of our conscious universe is where God will be found. However, some fortunate people have received glimpses of this universal mind and this fundamentally changes their lives for the better. Hopefully at some future point we will all experience this awareness. Our current knowledge of reality is fairly limited because creation is vastly more complicated than we conventionally understand. We must change the modern acceptance of this material world being synonymous with absolute reality. The evidence for life after death is overwhelming and the hypothesis here is meant to be more than just idle speculation. But there does seem to be at present, a deliberate almost Cosmic system in place, to keep our knowledge of what comes next uncertain and secretive. The afterlife is an objective reality that exists whether we like it or not and our passage there after physical death is an automatic process in spite of what many religious groups may claim. I have attempted here to set out a possible location for the next existence which is currently so well hidden and mysterious. It is clearly not anywhere within our understanding of normal space and time yet appears to be all around us. For this model of a (currently) hypothetical afterlife existence to work, only one further spatial dimension is required and we already have it: disguised as time. We will all leave this material existence one day and not knowing what happens next is a woefully negligent position to be in, yet there are too many clues to say there is nothing. When a meaningful afterlife is confirmed the changes to humanity will be astounding; the knowledge that we all have an ultimate purpose will transform human existence. We will acquire an awareness that our lifetime of choices and actions, while existing on the Earth's physical plane, have temporarily hidden long term consequences on the destination of our metaphysical bodies. Reassuringly, crime worldwide is likely to plummet. The extreme grief we experience on the death of people close to us will disappear when we realise they are actually still there and will ultimately be reunited with us.

Bibliography

1. Dawkins, Richard. The God Delusion, 2009
2. Sheldrake, Rupert. The Science Delusion: Freeing the Spirit of Enquiry. 2012.
3. Di Valentino, E. Planck Evidence for a Closed Universe and a Possible Crisis for Cosmology. Nature Astronomy, 4, 2019.
4. Gastanaga, E, et al, Gravitational Bounce from the Quantum Exclusion Principle. Journal of Physical Review D 111. 2025.
5. Hawking, Stephen. A Brief History of Time. 1998.
6. Wilczynska, M.R. et al. Scientific Advances, vol. 6 (17), 2020.
7. Granville, William. The Fourth Dimension and the Bible, 1922.
8. Abbot, Edwin. Flatland: A Romance of Many Dimensions. 1884.
9. Laszlo, Ervin. The Self Actualising Cosmos. 2014.
10. Ferris, Timothy. The Whole Shebang. 1998.
11. McDougall, Duncan. Hypothesis Concerning Soul Substance Together with Experimental Evidence of the Existence of Such Substance. American Medicine. 1907.
12. Martin, William Fergus. Afterlife Adventures: Life After Death Stories. 2023.
13. Moody, Raymond, A. Life After Life. 2001.
14. Monroe, Robert. The Ultimate Journey, 2014.

15. Reccia, Michael G. The Joseph Communications: The Fall. 2012.
16. Newton, Michael. Journey of Souls. 2010.
17. Cannon, Dolores. The Convoluted Universe. 2001.
18. Monroe, Robert. A Summary of Robert A Monroe's Journeys Out of the Body. 2022.
19. Ziewe, Jurgen. Multi-Dimensional Man. 2008.
20. Beard, Paul. Living On: How Consciousness Continues and Evolves After Death. 2015.
21. Lorimer, David. Resonant Mind and Life Review in the Near- Death Experience. 2017.
22. Cramer, Lynda. Five Years in Heaven. 2021.
23. Warren, Joshua. Pet Ghosts: Animal Encounters from Beyond the Grave. 2006.
24. Gordon, Stan. Silent Invasion: The Pennsylvania UFO-Bigfoot Casebook. 2010.
25. Peake, Anthony. The Hidden Universe. An Investigation into Non-Human Intelligences. 2019.
26. Geiger, John. The Third Man Factor: Surviving the Impossible. 2009.
27. Sheldrake, Rupert. A New Science of Life: The Hypothesis of Morphic Resonance. First published 1981.
28. Jung, C.J. Synchronicity: An Acausal Connecting Principle. 1960.
29. Streiber, Whitley. Communion. 1989.
30. Carroll, Judy. ZETA Message. Connecting All Beings in Oneness. 2010.
31. Steiger, Brad and Sherry. Real Aliens, Space Beings and Creatures from Other Worlds. 2011.
32. LaViolette, Paul. The Secrets of Anti-Gravity Propulsion. 2008.

33. Hennenberg, M. Decrease of Human Skull Size in the Holocene. Human Biology, 60, 395-405. 1988.
34. Bratsberg, B, et al. Flynn Effect and its Reversal are both Environmentally Caused. PNAS, 115 (26) 6674-6678. 2018.
35. Jansen, P. R. et al, Genome-wide meta-analysis of brain volume identifies genomic loci and genes shared with intelligence. Nature Communications, 11, article 5606. 2020.
36. Schnabel, Jim. The Secret History of America's Psychic Spies. 2011.
37. Dolan, B. and Zabel, R.M. AD: After Disclosure: The People's Guide to Life After Contact. 2012.
38. Greer, Steven. Hidden Truth Forbidden Knowledge. 2013.
39. Callaway, E. Oldest Homo sapiens fossil claim rewrites our species' history. Nature, June 7, 2017.
40. Foerster, Brien. Elongated Skulls of Peru and Bolivia: The Path of Viracocha. 2015.
41. Tompkins, P and Bird, C. The Secret Life of Plants: A Fascinating Account of the Physical, Emotional and Spiritual Relations Between Plants and Men. 2018.

www.ingramcontent.com/pod-product-compliance
Lightning Source LLC
Chambersburg PA
CBHW052054070526
44584CB00017B/2175